TECHNOLOGY SERIES   Technical Memorandum No. 5

I0036050

Small-scale oil extraction from groundnuts and copra

Prepared under the joint auspices of the International Labour Office and the United Nations Industrial Development Organisation

International Labour Office   Geneva

ILO
UNIDO
Small-scale oil extraction from groundnuts and copra
Geneva, International Labour Office, 1983. Technology Series, Technical Memorandum No. 5

/Technical memorandum/, /Small-scale industry/, /Agriproduct processing/, /Vegetable oil/, /Nut/s, /Developing country/s. 08.06.2
ISBN 92-2-103503-4
ISSN 0252-2004

ILO Cataloguing in Publication Data

CONTENTS

APPENDICES

ACKNOWLEDGEMENTS

The publication of this memorandum has been made possible by a grant from the Overseas Development Administration of the United Kingdom, through the Intermediate Technology Development Group (ITDG, London). The International Labour Office and the United Nations Industrial Development Organisation acknowledge this generous support.

Note to the UNIDO Edition

The choice of the most appropriate technology to be applied in
industrial production activities is one of the many problems which developing
countries face in the process of promoting industries in their countries.
An appropriate choice can only be positively made when there is an effective
and functioning flow of information about the available alternatives.

The International Labour Office (ILO) and the United Nations Industrial
Development Organization (UNIDO) are both engaged in activities to assist the
developing countries in the promotion of small and medium scale industries.
The two organizations agreed upon to develop joint programmes that could
contribute to enhancing the capacity for appropriate choice of technology.
One such programme was to make combined efforts in preparing a series of
Technical Memoranda in selected critical and priority sectors of industry in
order to disseminate information on alternative production technologies.

The present is the fourth volume in the series. The first, second and
third volume are entitled:
- "Tanning of Hides and Skins" (UNIDO/IS.326);
- "Small-scale Manufacture of Footwear" (UNIDO/IS.354); and
- "Small-scale Weaving" (UNIDO/IS.454).

It is hoped that the publication will be found useful in stimulating
the development of the small-scale oil extraction industry in the developing
countries.

G.S. Gouri
Director
Division for Industrial Studies

PREFACE

Food processing constitutes a major economic sector in developing countries for the following reasons. Firstly, processed food constitutes one of the most important basic needs goods, especially in urban areas where low-income families are not equipped to carry out the basic processing of agricultural and animal products. Secondly, food-processing allows the consumption of seasonal agricultural products over the whole year and therefore minimises the important price fluctuations resulting from the periodic gluts and shortages of the fresh products. Consequently, both the farmers and consumers benefit from a greater price stability. Thirdly, food-processing could generate substantial foreign exchange in countries which produce a large surplus of agricultural products.

Given the importance of the food-processing sector and its potential contribution to the achievement of major socio-economic objectives, the expansion of this sector should be carefully planned with a view to maximising its impact on the national economy. In particular, the choice of processing technology and scale of production should be such as to ensure the fulfilment of various development objectives such as the generation of productive employment, the production of low-priced processed food products suitable for low-income groups, rural industrialisation, foreign exchange savings, the generation of backward and forward linkages with other sectors of the economy, etc.

A number of developing countries recognise the importance of the food-processing sector and have developed and promoted food-processing techniques consonant with the country's development objectives. In particular, they have maintained an adequate balance between small-scale food-processing units using labour-intensive or intermediate techniques and large-scale units using imported, capital-intensive technologies.

Unfortunately, a large number of countries have not been able to maintain such a balance for various reasons, one of them being the lack of technical and economic information on alternative food-processing technologies. Thus, small food-processing units are being increasingly replaced by imported large-scale plants - often turn-key factories - which are not always suitable to local socio-economic conditions. The choice of inappropriate technologies has often led governments to subsidise large-scale food-processing plants or to supply them in priority with the needed raw materials to the detriment of the existing small processing units. Yet, these latter units could use improved technologies which have already been successfully adopted in a number of countries, and be more competitive than the imported large-scale plants. As detailed information on these technologies is not generally available to small-scale producers in developing countries, the International Labour Office has started a new series of technical memoranda on specific products and processes for wide dissemination among these producers as well as public planners and industrial development agencies which have an important role to play in the promotion of appropriate food-processing technologies. Eight technical memoranda on these technologies have already been published or are under preparation.[1] Some of these memoranda are being jointly published with the United Nations Industrial Development Organisation (UNIDO) and/or the Food and Agricultural Organisation (FAO).

This technical memorandum on oil extraction from groundnuts and copra covers the pre-processing of raw materials, oil extraction through pressing and the post-treatment stages. It provides detailed technical and economic information on small-scale oil extraction mills using either small expellers or power ghanis, and processing between 100 tonnes to 220 tonnes of raw materials per year. An economic comparison between these small-scale plants and medium- to large-scale plants is provided in Chapter IV. However, as this comparison is based on a number of assumptions regarding the market prices of raw materials (groundnuts and copra) and oil, equipment costs, etc., it should not be used as a basis for the choice of oil extraction technology and scale of production. Instead, the interested reader should undertake his own evaluation of the latter on the basis of the technical information contained in Chapters II and III, and local factor prices (e.g. wages, raw materials prices, oil prices).

As public planners may be interested in the socio-economic impact of alternative oil extraction technologies, Chapter V provides information on the following effects of small-scale and large-scale oil extraction units: employment generation, basic needs satisfaction, energy requirements, transport costs, and multiplier effects. It also suggests a few policy measures for the promotion of the right mix of oil extraction technologies.

---

[1] A technical memorandum on small-scale fish processing has already been published. Six technical memoranda on food-processing - in addition to this one - are at various stages of preparation. They cover the most important food products of interest to developing countries.

Whenever possible, detailed information is provided for the local manufacture of ancillary equipment which may be needed by small-scale oil mills. In addition, a list of equipment suppliers is included in an appendix of this memorandum with a view to facilitating the procurement of equipment (e.g. expellers, power ghani units, filtering equipment) by small-scale producers. It must, however, be emphasised that the supply of names of equipment manufacturers does not imply a special endorsement of the latter by the ILO. These names are only provided for illustrative purposes, and oil producers should try to obtain information from as many suppliers/manufacturers as feasible.

This memorandum does not describe all existing oil extraction technologies. Rather, a choice has been made from among those which have been successfully applied by small-scale millers in a number of developing countries. Other technologies, not described in this memorandum, may also be adapted to local conditions and tried in a few production units with a view to assessing their technical and economic efficiency. The bibliography in Appendix VI should provide useful information on these technologies.

A questionnaire is attached at the end of the memorandum for those readers who may wish to send to the ILO or UNIDO their comments and observations on the content and usefulness of this publication. These will be taken into consideration in the future preparation of additional technical memoranda.

This memorandum was prepared by J. Keddie (consultant) and M. Allal, staff member in charge of the series of technical memoranda within the Technology and Employment Branch of the ILO. Mr. A.M. Das (consultant) collaborated in the economic evaluation of alternative oil extraction technologies.

A. S. Bhalla,
Chief,
Technology and Employment Branch.

CHAPTER I

INTRODUCTION

The extraction of vegetable oils from oilseeds is a well-established industrial activity in a number of developing countries. This is due to the fact that, since the early 1950s, most oilseed-growing countries have favoured indigenous oil extraction in preference to the export of oilseeds. They have thus supported the setting up of factories for this purpose, which, to a great extent, are large-scale plants situated in or near urban areas.

I.1   Scope of the memorandum

This memorandum is primarily concerned with the choice of technology for the extraction of unrefined oil from groundnut kernels and copra by small-scale mills located in rural or small urban areas. This follows from the fact that a great majority of the population in developing countries reside in rural and small urban areas, and that the techniques they use for oil extraction offer substantial scope for improvement. Another important reason is that the extraction of unrefined oil at small-scale levels does not require sophisticated technology.

The inputs of raw material associated with the terms small-scale, medium-scale and large-scale used in this memorandum are as follows

(i)   small-scale plant: 350-800 kg of raw material per day;

(ii)   medium-scale plant: 8,000-20,000 kg of raw material per day;

(iii)   large-scale plant: 25,000-100,000 of raw material per day.

I.2   Oilseeds considered in the memorandum

There are a number of types of oilseeds grown in developing countries. This memorandum, however, considers only the extraction of groundnut oil from groundnut kernels and that of coconut oil from copra. A number of factors led to this choice, including the following: a significant indigenous consumption of groundnut and coconut oils in all the oilseed-producing countries; a substantial demand for unrefined oil in these countries; and the relatively high oil content of coconuts and groundnuts.

I.3   Technologies covered by the memorandum and target audience

The memorandum considers the use of expellers at small-scale levels as well as that of power ghani units (small-scale).[1] The oil extraction techniques associated with these two types of equipment are described in sufficient detail to allow their application by small-scale entrepreneurs. An economic analysis of both small-scale and large-scale production units is also undertaken in order to enable interested government officials to compare the socio-economic impact of alternative oil extraction technologies.

The organisation of the chapter on processing techniques (Chapter III) describes the processing associated with each technique as well as other aspects, such as daily usages, labour requirements, equipment schedules and a recommended floor plan.

The memorandum is primarily intended for entrepreneurs who either wish to upgrade an existing oil plant (in terms of efficiency), or are considering to start a small-scale oil-extraction unit. It should also be of interest to smallholders who may wish to process their produce, either groundnuts or coconuts, into the respective oil.

I.4   Summary of remaining chapters

Chapter II considers briefly the raw materials covered by this memorandum, namely groundnuts and coconuts. It then proceeds, in greater detail, to outline the pre-processing stages involved with respect to these materials. In particular, the chapter reviews a number of techniques used in developing countries to dry copra. It may be noted that in most cases, the processing of fresh coconuts into copra is undertaken at the farm rather than at the oil-extraction plants. On the other hand, the shelling of groundnuts is often carried out at the plant as shelled nuts tend to spoil faster than non-shelled nuts.

Chapter III describes in detail the various oil extraction technologies, while Chapter IV analyses the economic efficiency of these technologies. Chapter V analyses the socio-economic implications of choosing a particular technique in preference to others.

---

[1] The power ghani mill is an improvement of the traditional ghani mill used in India for the crushing of oil seeds. For detailed information on this type of mills, see Chapter III.

CHAPTER II

PRE-PROCESSING OF COCONUTS AND GROUNDNUTS

Prior to describing the various oil-extraction techniques, it is necessary to consider the raw materials involved (i.e. copra and groundnuts) and the pre-processing operations. However, as this memorandum is mostly concerned with oil extraction, information on these operations will be limited. The reader is therefore advised to obtain additional information from available publications, some of which are indicated in the chapter and in the bibliography.

II.1   Raw materials

The two raw materials of concern to this memorandum are copra and groundnuts. Copra is derived from coconuts. The derivation process is dealt with in the next section.

Coconuts and groundnuts are produced in developing countries mainly by smallholders. Table II.1 provides certain basic coefficients involved in the cultivation of coconuts and groundnuts, including the proportions of coir fibre, shells, coconut water, etc. per tonne of crude nuts.

III.1.1 Copra

Copra is available in two forms, namely, cup copra and ball copra. The preparation of cup copra is discussed in detail in the next section. Ball copra is obtained from mature unhusked nuts that have been stored in the shade for eight to 12 months. The nut water is gradually absorbed and the kernel dries up until it can be heard to rattle inside the shell when the nut is shaken. Husk and shell are then carefully removed. This copra is used for edible purposes. Table II.2 provides the composition of the kernel from both mature coconuts and copra.

III.1.2 Groundnuts

Groundnuts are grown in tropical and subtropical climatic regions and warmer parts of temperate regions. It is a low-growing annual plant, between 30-60 cm high. The varieties of groundnuts fall into two groups depending on

Table II.1

Basic coefficients[1]

	Coconuts	Groundnuts
Nuts	14 000/ha[2]	N/A
Nuts/tonne	1 000	N/A
Tonnes/ha (crude nuts)	15	1 (higher with irrigation)[3]
Shelled nuts/ha (Fresh)	5 250 kg	700 kg
Shelled nuts/ha (dry)	2 550 kg	670 kg
Shelled nuts/tonne crude nuts (dry)	170 kg	670 kg
Shelled nuts/tonne crude nuts (fresh)	350 kg	700 kg
Coir fibre/tonne crude nuts	150 kg	N/A
Coconut water/tonne crude nuts	140 kg	N/A
Shells/tonne crude nuts	150 kg	300 kg
Shells charcoal/tonne crude nuts	45 kg	40 kg
Nuts in shell/tonne crude nuts	640 kg	1 000 kg
Coir waste/tonne crude nuts	210 kg	N/A

[1] The basic coefficients vary a great deal, depending on soil characteristics, growing conditions, rainfall, type of nut, etc.

[2] This is a high estimate corresponding to ideal conditions, with approximately 140 palm trees per ha and 100 nuts per tree. Under unfavourable conditions, the yield may be as low as 1,500 nuts per ha.

[3] The yield varies greatly according to location and growing conditions. Obtained estimates range from 500 kg/ha (Africa) to 2 250 kg/ha (United States).

Table II.2

Composition of mature coconuts and copra[1]

	Percentage (wt) coconut kernel	Percentage (wt) copra
Oil	35-40 (on a wet basis)	60-70
Protein	4-5	8-10 (crude protein)
Moisture	40-50	5-7
Carbohydrate	5-10	10-13
Crude Fibre	2-4	4-5
Ash	1-1.5	1-2

[1] Variable composition of kernels and copra depending on origin of fresh nuts and pre-processing conditions.

their characteristics:

> (i) Bushy bunched types: this type matures in 3-4 months.
>
> (ii) Runner or spreading type: this type matures in 4-6 months.

Some intermediate hybrids do exist. The bunch type contains kernels that average 65-75 per cent of the whole nut. A good average yield, under suitable conditions, would be 1000-1350 kg/ha.

The major groundnut-producing countries are India, China, Nigeria, Senegal and the United States. While the major proportion of groundnuts is used for the production of vegetable oil, a fairly substantial amount of shelled nuts are used for direct human consumption either raw or roasted.

II.2   Pre-processing of coconuts

The pre-processing operations of coconuts are indicated in the flow chart below, and then briefly described in this section.

> Harvesting   Husking   Shelling   Drying   Bagging   Storage

II.2.1   Harvesting

Harvesting of coconuts can be achieved by two means: (i) allowing the nuts to fall naturally or, (ii) causing them to fall artificially. In the latter case, they may be picked manually or with a knife attached to a long bamboo pole. To obtain good quality copra, the nuts must be harvested fully ripe. A mature bunch of coconuts is usually ready for harvest every month. Therefore, ten to 12 harvests may be obtained each year.

II.2.2   Husking

(i) Manual husking

The first stage of coconut processing is the removal of the husk. The most prevalent method is a manual one. It involves the removing of the husk by impaling the coconut on a sharp iron or wooden spike fixed in the ground. The spike is usually set at a slight angle at about 80 cm from the ground. Using both hands, the operator brings the coconut down sharply on the edge of the spike which pierces the husk and glances off the round end of the nut. The fruit is then twisted, thus loosening the husk. This work is hard, and calls for considerable skill and great wrist and arm strength. An experienced worker can dehusk approximately 1500-2000 units per eight-hour day. Figure II.1 shows two typical manual dehusking tools.

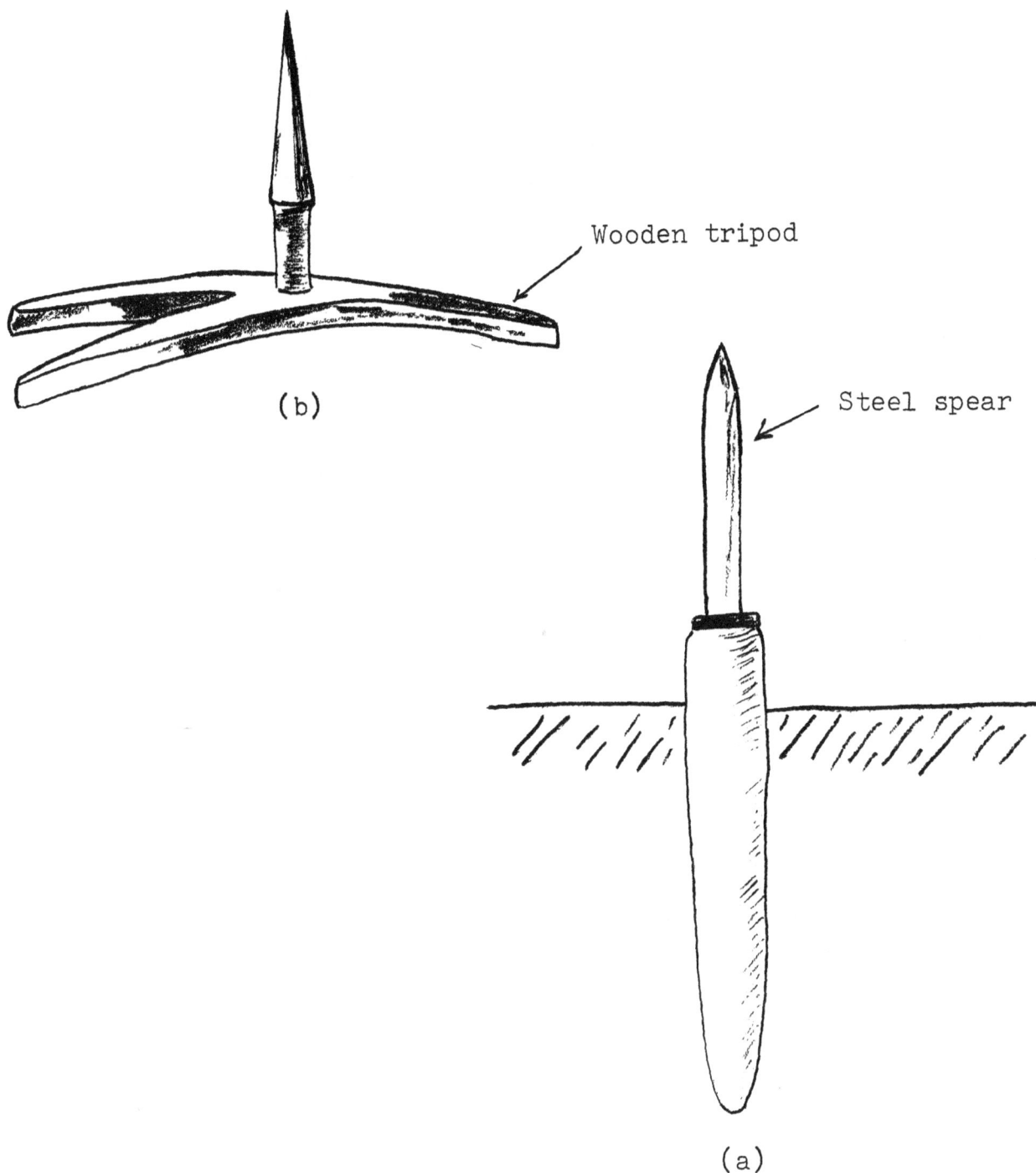

Wooden tripod

(b)

Steel spear

(a)

FIGURE II.1

Two types of manual dehusking tools

(a)   This dehusking tool is used in Malaysia and
Sri Lanka. The wooden base is driven into the ground.
The upper spear is made out of steel.

(b)   In other developing regions, a coulter mounted
on a wooden tripod is used for husking.

(ii) Mechanical husking

In view of the rigour involved in husking by hand, a number of machines to husk coconuts mechanically have been devised. Up to now, none of these machines have been able to compete satisfactorily with the traditional method described above. However, in view of the ever increasing labour costs, a good deal of research is being conducted to evolve suitable equipment for mechanical husking.

II.2.3 Shelling

There are various techniques for shelling coconuts depending on the overall processing method. To produce copra, the shell of the dehusked coconut is cracked into two equal halves with a chopping knife or hatchet. Breaking should be carried out over a concrete floor sloping either to one side or to the centre so as to enable the draining of coconut water. A skilled worker can split up to 7000 nuts per day. The nut water should be drained completely by inverting the split nuts face downward for one to two hours, preferably in the sun, before the drying is started.

Once drainage has been completed, drying should be undertaken as soon as possible in order to avoid bacterial action and spoilage of the meat. The latter generally starts approximately four hours after shelling. Experiments on the rate of spoilage were undertaken in Malaysia. Estimated spoilage rates are shown in Table II.3. It may be seen that the percentage of white copra decreases from 82 per cent (no delay between shelling and drying) to 0 per cent after a 48-hour delay.

II.2.4 Drying

There are three main drying methods. These are:
- sun drying;
- smoke curing or drying over an open fre in a direct drier or kiln; and
- indirect hot-air drying.

Frequently, a combination of preliminary sun drying followed by kiln drying is used. The selection of an appropriate technique would, of course, be largely influenced by local conditions. A recommended drying procedure would proceed as follows:

Table II.3

Effect of preliminary drying delays on the quality of copra

Experiment number (300 nuts per trial)	Delay period before kiln drying	Percentage white meat	Percentage slightly dis-coloured or dirty meat	Percentage red to red-black meat
1	nil	82	18	0
2	2 hours	80	20	0
3	4 hours	75	25	0
4	6 hours	70	29	1
5	9 hours	61	35	4
6	12 hours	36	42	22
7	24 hours	10	48	42
8	48 hours	0	17	83

Source: Thampan, P.K.: The coconut palm and its products (Cochin, India, Green Villa Publishing House, 1975).

- The moisture content is first reduced from 55 per cent to 35 per cent within 24 hours;

- Over the next 24 hours the moisture content should be further reduced to 20 per cent;

- In the third 24-hour period the moisture content should be reduced to 5-6 per cent.

(a) Sun drying

The sun-drying process is very simple but is successful only when long periods of sunshine can be guaranteed. The process involves placing drained coconut halves on the ground with the open side turned to the sun. It is better, however, to first immerse the nut halves in clean, cold water to wash away any adhering material. This is done so as to prevent those materials from acting, later on, as a medium for the development of mould.

The nuts may be dried on simple and cheap racks, constructed of bamboo, in order to avoid contamination by laying them directly on the ground. Technically, the use of concrete floors is superior to both of the above methods. These will, however, require higher capital investments which may not always be justified from an economic point of view.

Depending on climatic conditions, the drying process may be completed within five to seven days. After the first two days of drying, the kernel or the meat gets detached from the shell and can be removed by means of a thin wooden lever. The detached kernels are set to dry again for a further period of four to five days. For best results, the drying period should not be discontinued. Table II.4 provides estimates of the moisture content of copra at different times of the drying cycle.

The drying area should be covered with screens of bamboo and coconut fronds whenever it rains in order to prevent deterioration of the partly dried meat.

The drying of coconut meat during the monsoon period is particularly difficult and, unless it is carried out carefully, a large fraction of the copra may get mouldy or discoloured. Table II.5 (based on experiments carried out in the Philippines) shows that the appearance of moulds is considerably retarded as the moisture content of copra is decreased.

TABLE II.4
Moisture content of copra during sun drying

Drying period	Percentage moisture	Drying period	Percentage moisture
0	50	2 days	17
2 hours	40	3 days	12
4 hours	35	4 days	9
5 hours	33	5 days	7
8 hours	28	6 days	6
10 hours	25		

Source: Thampan, op. cit.

TABLE II.5
Relationship between moisture content of copra and mould appearance

Drying period	Average temperature	Average percentage moisture (wet)	Mould appears after
Fresh meat	0	53	3 days
1 day	28.1°C	26	7 days
2 days	28.4°C	20	8 days
3 days	28.9°C	10	13 days
4 days	29.0°C	8	28 days
5 days	29.0°C	4	---

Source: Thampan, op. cit.

(b) <u>Direct hot-air drying kiln</u>

In this form of drying, coconut meat is heated directly by the smoke and hot combustion of fuel gases. Crude kilns used for this type of drying yield an inferior, heavily smoked, product. Consequently, .the produced oil and cake are also of poor quality. There are, however, certain kilns that can be employed very satisfactorily. An FAO Agricultural Development Paper[1] considers a few direct and indirect hot-air kilns. One of each type of the above kilns is described in this section.

The FAO paper provides the following description of the direct hot-air drying kiln:[2]

This miniature kiln, designed in Malaysia for use on smallholding varying in size between 0.7 and 4 ha is able to dry 100 coconuts and yield crisp, white, undistorted, and uniformly dried copra in fifteen hours ('See fig. II.2).

High ground is selected for the site of the kiln, which is a rectangular wooden chamber 1.8 m high and open at the top. A tunnel leading into the base of this chamber forms the hearth in which a single chain of coconut shells is burnt continuously. The overall area of chamber and tunnel is 2.1 x 0.9 m, and it is sheltered by an attap roof.

In the chamber the meat is dried on a platform of spaced nibong slats (the palm "Oncosperma filamentosum") covered with chicken wire and placed horizontally within the chamber, 1.5 m above ground level. About 7-8 cm beneath this platform there is a heat-spreader, supported horizontally by wire. This piece of iron is perforated with holes. The base of the chamber is filled to a height of about 0.5 m with rammed clay, leaving a hole in the centre which is lined with loose bricks, through which the hot gases emerge from the tunnel. A drop shutter is provided at the top of the chamber, so that the meat may be easily inserted or removed and the heat spreader may be cleaned when necessary. There is also an inspection door about 45 cm above ground level for cleaning the clay base of the chamber.

---

[1] FAO <u>Coconut palm products - their processing in developing countries</u>, FAO Agricultural Development Paper No. 99 (Rome, 1975).
[2] ibid., pp. 59-61.

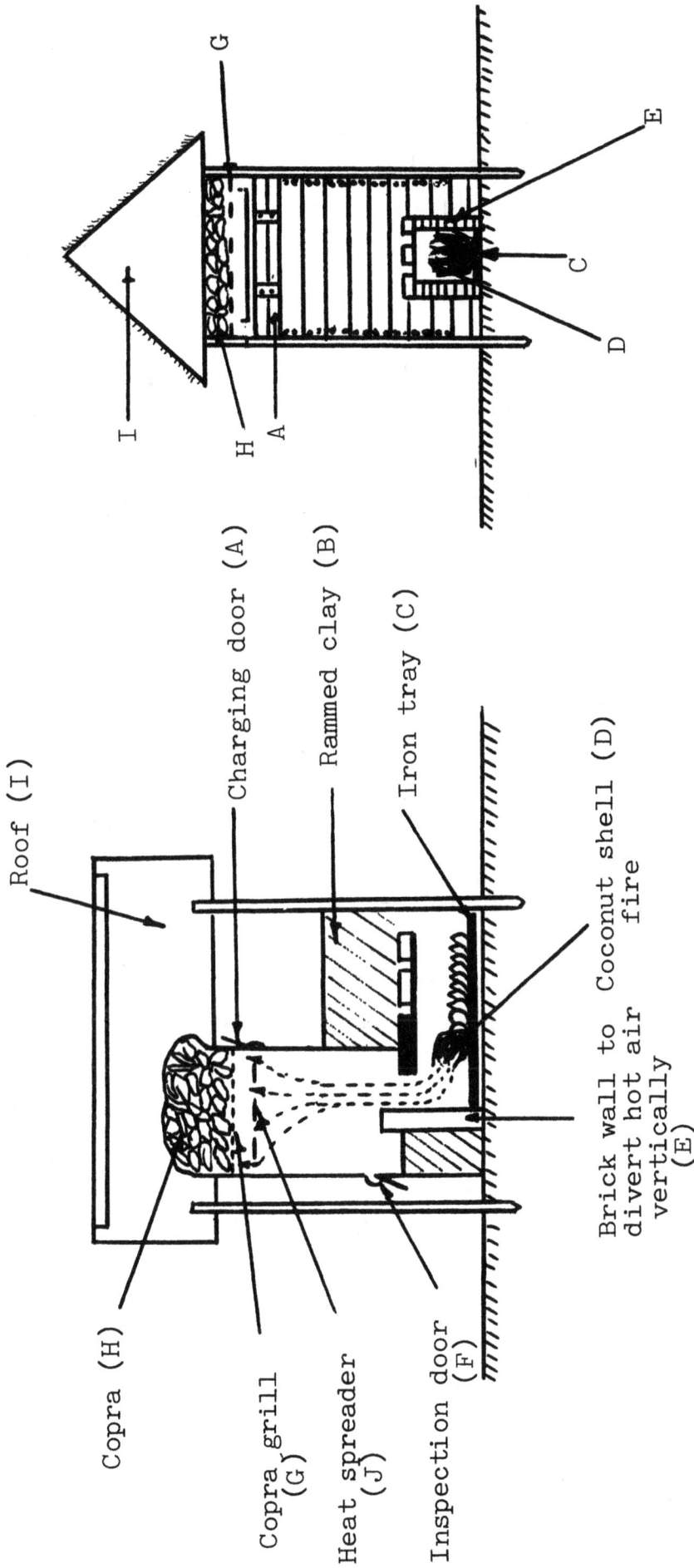

FRONT VIEW

CROSS SECTION

Roof (I)

Charging door (A)

Rammed clay (B)

Iron tray (C)

Brick wall to divert hot air vertically (E)

Coconut shell fire

Copra (H)

Copra grill (G)

Heat spreader (J)

Inspection door (F)

Figure II.2

MALAYSIAN "TEN-ACRE KILN" (COOKE TYPE)

The fire tunnel consists of two walls of loose bricks laid along the ground and leaning into the base of the chamber. These walls, each six bricks high, are laid 30 cm apart, and the space between them is covered by a stout sheet of iron, 1.8 m long and 46 cm wide, anchored in position by bricks laid along the edges and by rammed clay where it joins the base of the kiln. The bottom line of bricks is spaced so as to allow air to filter to the fire.

To operate the kiln, clean, dry coconut shells of even size are interlocked and laid loosely in a single line on a sheet iron tray, 1.5 m long and about 20 cm wide, outside the kiln. The hollow end of the row is lit, using scrap rubber or kerosene to assist combustion. When the shells are well lit and burning without smoke, the tray is slid as far into the tunnel as it will go, the burning end entering first. Thus, the shells burn from the centre outward against the incoming current of air. This keeps the flame small and reduces the rate of combustion. If the shells were burned in the opposite direction, the draught would cause all of them to ignite simultaneously, thus spoiling the meat and possibly also destroying the kiln. It should be noted that if the tunnel is too small in cross section, the induced draught is so great that the fire burns with difficulty and gives a smoky flame.

When the fire is nearly out and beginning to smoke, the tray is withdrawn and reloaded with a fresh supply of shell fuel, and just before this second fire is finished, the meat is turned over. Subsequently, three more fires are lit in succession until the meat is dry and ready for shelling.

The following can be said about the efficiency of this type. A single line of shells 1.5 m long is made up of about 40 pieces (halves) of shell and lasts about three hours. Thus a total of 200 pieces are used to fuel the kiln five times, which means that all the shells obtained from one lot of 100 coconuts are consumed in the subsequent run. It follows that with such rapid drying the heat efficiency of the kiln tends to be low, but this does not matter so long as sufficient fuel is produced to meet the requirements of firing.

The drying time is remarkably short, and although the fires need fairly frequent attention when in operation, intervals may safely be left between fires when the copra is nearly dry without causing deterioration. Once a supply of dry shell fuel is available and the kiln and ground have dried out, copra of estate quality should be obtained irrespective of weather conditions. Properly operated there is little fire risk. If the poles and woodwork are initially and periodically treated with creosote, a life of over two years may reasonably be anticipated, but the attap should be renewed yearly.

In a year of 150 working days, this kiln will produce about 3700 kg of copra. When the kiln is shared by two or more smallholders, it can be operated more economically.

A modified Malaysian kiln has been developed in India. It is particularly recommended for small coconut holdings. The kiln can dry 200 nuts (400 cups) at a time in approximately 34 hours. It is relatively inexpensive and may be afforded by most producers. Except for a slight smoky smell, good quality copra can be produced from this kiln.

(c) Indirect hot-air drying

In indirect hot-air driers the coconut meat does not come into contact with combustion gases and smoke from the fuel and hence the name of this drying method. The copra thus obtained is usually of a very high quality. However, unless the method of heat exchange is efficient, there is a considerable loss of heat.

The FAO paper mentioned earlier provides a description of an indirect hot-air drier known as the modified Tonga hot-air drier. These driers are cheap and easy to build. They enable pre-drying of copra in the husk with hot air, while the upper trays of the drying chamber may be used to dry cut-out copra. Thus, pre-drying is no longer dependent on the sun. Following is a description of these driers as provided in the FAO paper:[1]

The flooring of the drying chamber is made of heavy 5-cm wire mesh (see figure II.3). After firing and when the drier is sufficiently hot the whole nuts are split in half and stacked in the chamber the first layer face up and the remainder face down.

---

[1] FAO, op. cit., pp. 85-88.

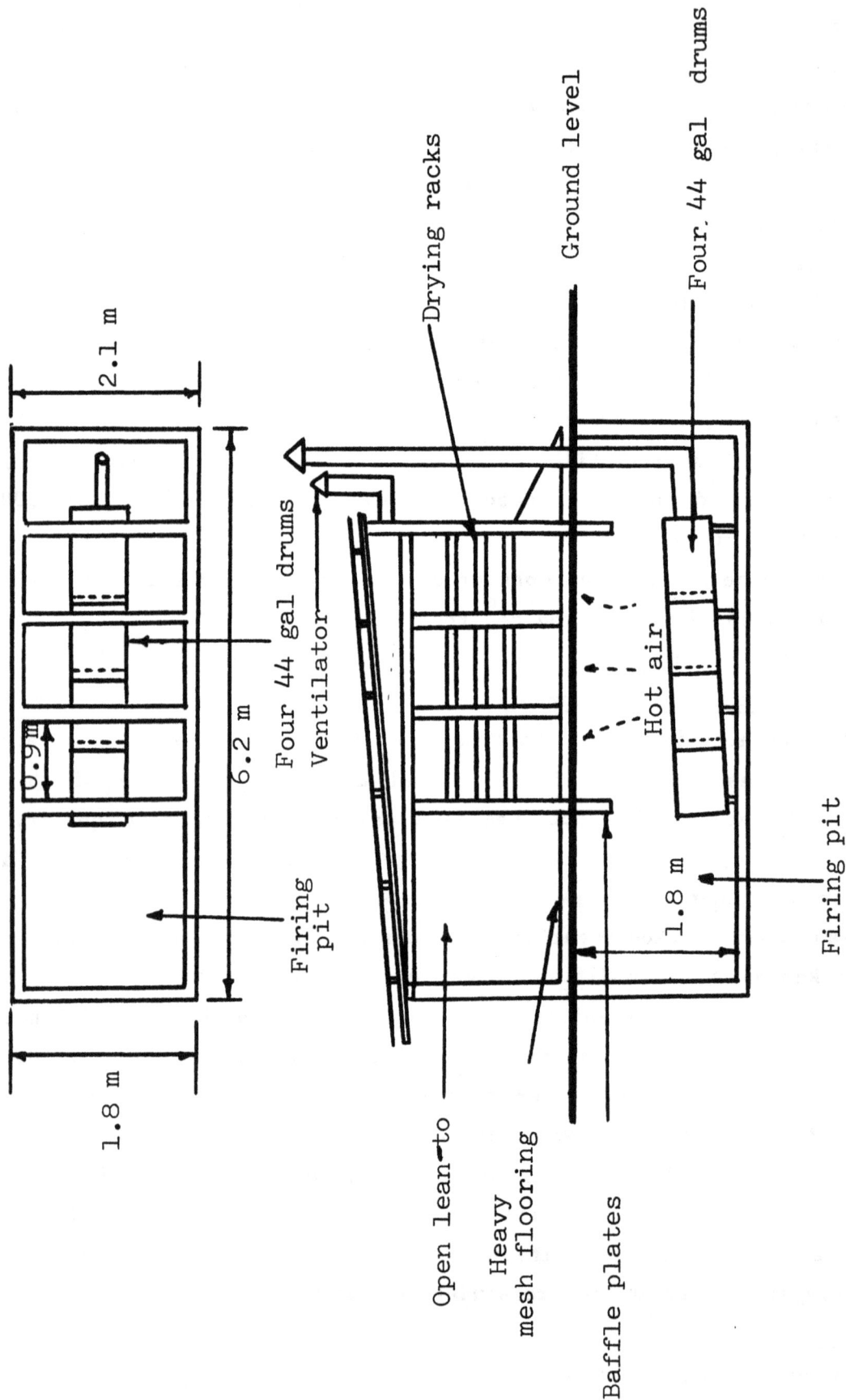

Figure II.3

TONGA HOT-AIR DRYER

Predrying is carried out for twelve hours. Next morning the half nuts are removed and the meat is cut out and put into the trays above for further drying. For conservation of shipping space it is important that the half nuts be cut into two or three pieces, but not more than four. Drying is usually done during daylight hours only - the furnace being stocked up in the morning and kept going all day, but left to go cold overnight. After this drying stage, usually for two days, the copra is taken out of the drier and spread out in the sun for two or three days more. A number of producers keep stoking at less heat after the first two days for a further day or two for complete drying. This saves final sun drying, particularly during bad weather.

Specifications

The pit is 1.8 m wide and 5.5 m long. The walls can be left bare, but it is preferable to use old sheet iron as lining to keep earth from falling into the pit.

Furnace and heating chamber

Four 44-gal drums are used, the ends of which are let in about 15-23 cm to make a single chamber. At ground level 10 x 5 cm local timbers are used as plates around the pit.

Drying chamber

The drying chamber is built of local timber directly over the heating chamber. The flooring at ground level is heavy 5 cm wire mesh, which will support half nuts in husk. The upper portion has three compartments, each with five racks carrying 1.25 mm^2 wire-mesh trays measuring 1.8 x 0.9 m - a total of fifteen trays in all. Trays can be made smaller (0.9 m square) with two trays in each rack for easier handling. Three hinged doors on one side are used for charging and discharging husks and trays. The walls and doors are made from empty drums.

The chamber has a capacity of 800 nuts per drying - that is, 150-250 kg of dry copra. The required drying times are twelve hours of predrying, twenty four hours of drying after cutting the copra out of the husks, and finally two days of sun drying or, alternatively, thirty-six hours of kiln drying.

Pattern of drying

As long as the drier is in operation the lower chamber of predrying in the husk can be filled and discharged while the upper trays are in use. Since normally the capacity of the top trays with cut-out copra will be about two or three times that of the lower chamber filled with husks, after discharging the predried copra the upper chamber should again be filled for predrying while the first batch of predried half nuts is being cut out and placed in the top trays. After the second batch has been predried and cut out, both batches are finally dried in the drying trays. The number of batches that are predried depends, of course, on the number of nuts to be dried.

The most important part of drying is the quick application of heat after the nuts are split. The furnace should therefore be stocked prior to splitting and the half nuts be put into the heated drying chamber as soon as possible.

Importance of sufficient heat

When predrying in the husk, the hot, moist air passing through the newly stacked husks will be fairly cool on reaching the cut-out, partly dried copra in the upper racks and will condense. This will not be the case, however, if the furnace has been fired and the heating chamber sufficiently heated prior to stacking the half nuts in the husk in the drier. It is therefore important to fire the furnace and heat the drier about an hour prior to stacking husks for predrying.

A number of indirect hot-air drying kilns have been developed in a number of countries, including the Pearson's patented kiln, the "kukum" hot-air drier, the "Chula copra drier", the Seychelles Calirifère, or the oil-fired hot-air drier marketed by M/S Premier Engineering Company at Cochin (India). These driers are particularly suited for specific scales of production ranging from approximately 500 nuts per load (the kukum drier) to 20,000 nuts per load (the Cochin drier). It is reported that the Chula copra drier (5,000 nuts/load/24 hours) produces copra of excellent quality which fetches a premium over ordinary copra.

The selection of a copra drier is a function of the adopted scale of production, price incentives for good quality copra and the financial means of the producer. Efficient driers are generally used by large-scale processors which can afford the high capital costs. On the other hand, small-scale producers tend to use traditional methods (e.g. sun drying in combination with direct hot-air drying) which are not always efficient from a quality point of view. Improved kilns have been promoted in a number of countries (e.g. India) but have not been widely adopted owing to the relatively high capital costs and the lack of price incentives for good quality copra. Thus, the widespread adoption of improved kilns by small-scale producers will require a differential pricing system for copra which benefits both the copra producer and the buyer (e.g. the middlemen or oil producer).

It may be noted that the quality of copra is determined by a number of factors in addition to the adopted drying technology. In particular, the oil content of copra and/or the oil extraction rate vary according to the variety of coconut processed (e.g. the copra of dwarf palms is soft and rubbery unlike that of the ordinary tall variety), its degree of ripeness, and growing conditions. The physical characteristics of copra which affect the oil quality and the extraction rate include the rubberiness of the meat, case hardening which results from inadequate drying in kilns and hot-air dryers, or charring of the meat from over-heating. Case hardening may be avoided by controlled drying of the copra cups: the temperature should not exceed 70°C during the early drying stage (e.g. the first eight hours of drying), and should then be maintained between 50°C and 60°C during the remaining drying stages. The charring of the meat should not happen either, if the above drying procedure were to be adhered to.

As the focus of this memorandum is oil extraction rather than the drying of oil seeds, no further detailed information will be provided on drying techniques. However, interested readers may obtain detailed information on a large number of copra driers used in a number of Asian countries from the following publication:

> UNIDO Coconut Processing Information Documents, Part 1 of 7, Coconut harvesting and copra production (Vienna, 1980), limited distribution.

The above publication provides a detailed description of 30 different driers used in Indonesia, Thailand, the Philippines, Malaysia, Micronesia, the Solomon Islands and Sri Lanka, including the design and operation of the kilns, construction methods, fuel and labour requirements, drying schedules, etc.

II.2.5 Bagging

Bagging is an optional operation. It is normally not done at small scales of production. If, however, bagging were to be done at this level, it would involve manual methods using jute bags.

II.2.6 Storage

Important points to be remembered with respect to the storage of copra are:

(i)   wet copra should, as far as possible, not be stored with dry copra;

(ii)   the storage structure should be constructed so as to minimise fluctuations in the storage climatic conditions. For example, a non-reflective metal roof which admits heat radiation to the store can be dramatically improved by painting the upper surface with a mat-white reflective paint. This can reduce temperature fluctuations in the store by $10^{o}C$ or more, thus preventing serious condensation effects.

(iii) The walls and floors should be smooth for easy cleaning. Cracks and crevices in the structure must be regularly cleaned out and filled with mortar to eliminate insect and rodent problems.

(iv)   Bagged copra should not be stored directly against walls; wooden dunnage should be provided to raise the stack off the floor.

Careful storage of copra will minimise the development of moulds which may lead to important losses of oil and an increase in the free fatty acid content of the meat. Five main types of moulds may develop depending on the relative humidity level and the temperature in the storage area. The most damaging mould is known as the "brown mould". It flourishes in copra with a moisture range of 8 per cent to 12 per cent and may be responsible for oil losses of more than 40 per cent. The least damaging mould is known as the "green mould": its growth is entirely superficial and does not result in significant oil losses.

In order to avoid the development of moulds, it is recommended to pre-dry copra in the sun for one to two days prior to storage with a view to reducing moisture to a safe 6 per cent to 7 per cent. This practice is usually adopted by a number of milling establishments in India. Relative humidity in the storage area should not exceed 85 per cent at room temperature or 95 per cent at 40°C.

II.3   Pre-processing of groundnuts

The pre-processing operations of groundnuts include, in the following order, harvesting, field drying, shelling, bagging and storage. These operations are briefly described below.

II.3.1   Harvesting

The timing of the peanut harvest is critical since it can greatly affect the yields and nut quality. As harvest time approaches, the peanuts should be inspected every day or two to determine the best date for digging. While the type, variety and planting date may be used as rough guides for determing the harvesting date, the best way to judge the maturity of the plant is to examine the pod itself. The crop is ready to harvest if the majority of the kernels are fully developed and take on a mature colour. Harvesting consists of either digging or pulling up the plants manually or using mechanised means such as a digger or southern plough.

II.3.2   Field drying

After the groundnuts have been harvested, they are inverted and placed in windrows in the field. The groundnuts are left to dry for about two weeks. During this period, the moisture content of the pods is reduced to about 10 per cent. In humid areas, the pods are sometimes picked off first and dried on mats so that they can be stacked or covered in the event of rain.

II.3.3   Shelling

Removing the kernels from the pods is generally referred to as shelling or "decorticating". This is usually carried out on the farm just before the farmer sells his produce for the following tow reasons: (i) kernels do not store as well as nuts in the shell and (ii) groundnuts in the shell are 50 per cent heavier than kernels alone and are therefore costlier to transport.

On smallholdings, groundnuts are shelled manually. This is a laborious and labour-intensive operation. Fortunately, a number of simple hand-operated decorticators are now available. These decorticators can be fitted with a simple feeder to improve the performance of hand-operated groundnut shellers.[1]

An improved decorticator involves the shelling of peanuts in a drum-shaped device with heavy, curved grates forming the lower half of the drum, and a revolving beater inside the drum which crushes the pods against the ridges in the grates. The clearance is sufficient to avoid injuring the peanut kernels when the shell is crushed. Peanuts and broken shells drop through the openings in the grates, and the shells are siphoned off by air suction. After the peanuts are shelled, the kernels are passed over oscillating shaker screens and separators where foreign material, undersize kernels, unshelled peanuts and split kernels are removed. On completion of this operation, the kernels are placed on a conveyor belt where defective kernels and any remaining foreign material can be removed by hand.

A number of groundnut shellers are available for various scales of production and powered by various means (manual, diesel engines, electric motors). Some of these shellers are illustrated and described in figure II.4.

Investment costs, energy costs, labour cots and productivity levels should be carefully analysed when choosing among alternative types of shellers. The repair and maintenance of the latter should also be taken into consideration as they do add to shelling costs and may disrupt shelling activities.

II.3.4 Bagging

This operation, as with copra bagging, is optional in small-scale production. If this stage is deemed necessary, manual methods of bagging using jute bags would be the most appropriate.

---

[1] A simple feeder made from steel is described in a Tropical Products Institute Publication (TPI: A feeder to improve the performance of a hand-operated groundnut sheller, Rural Technology Guide No. 4 (London, 1977)). The feeder consists of a large hopper with a cup valve in the base which drops a controlled quantity of nuts into the ball of the machine with each movement of the operating handle. Thus, as the number of nuts in the machine is kept small, the effort to operate the machine is lower than normal. The use of a hopper also results in less damage to the individual kernels.

GROUNDNUT SHELLER

Manufactured by:

    Agrimal (Malawi) Ltd.,
    P.O. Box 143,
    Blantyre,
    Malawi

The Agrimal groundnut sheller is a reciprocating decorticator equipped with inter-changeable screens for groundnuts of different size.

Price : (as at 1.4.80) 105.00 kwacha
                     ($128.00)

Source: Commonwealth Secretariat: Guide to technology transfer in East, Central and Southern Africa.

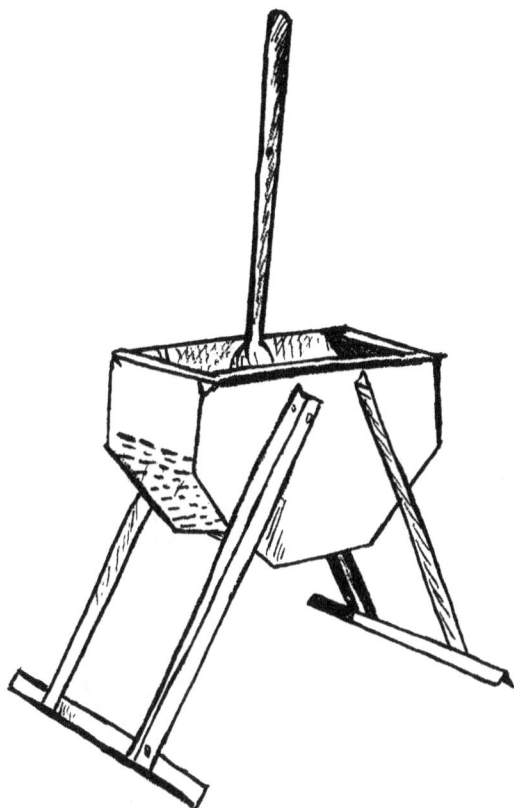

Fig. II.4 (a)

FOOT OPERATED GROUNDNUT SHELLER

Manufactured by:

    Hindsons Pvt. Ltd,
    The Lower Mall, Patiala,
    Punjab,
    India

Fitted with a flywheel for easier operation and with a blower to separate the shells from the kernels.

The machine can be operated and fed by one person and can shell 200 kg in an eight-hour day.

Source: ITDG: A buyer's guide to low-cost agricultural implements.

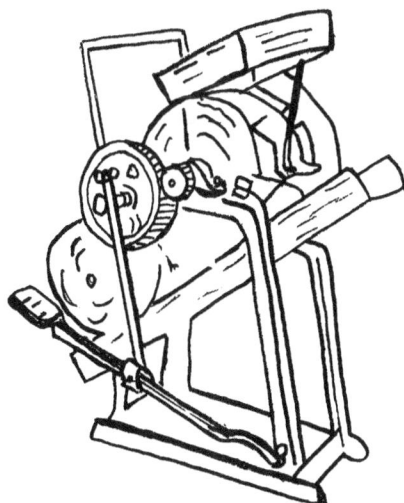

Fig. II.4 (b)

FIGURE II.4

Different types of groundnut shellers

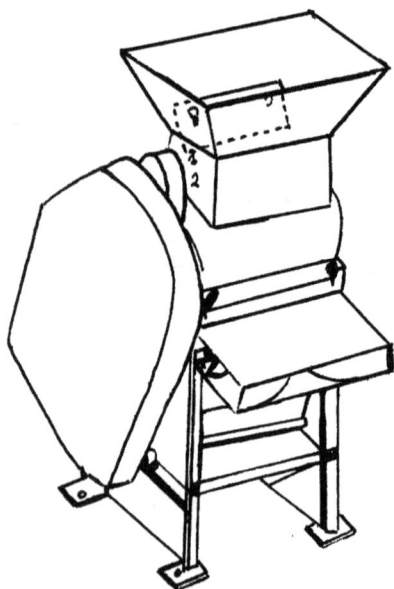

Fig. II.4 (c)

AUTOMATIC GROUNDNUT DECORTICATING MACHINES

Manufactured by:

Harrap, Wilkinson Ltd.,
North Phoebe Street,
Salford M5 4EA,
United Kingdom

Operation

The hopper at the top of the machine is filled with nuts to be shelled. A ribbed feed roller feeds the nuts into the beater chamber where they are struck by rotating flexible beaters. The broken shells and the kernels are forced out through the perforated cylindrical steel shelling screen (available in three sizes). The kernels and broken shells fall into a duct which has a wire-mesh delivery chute at its lower end. A fan blows the shells upward and out of the shell outlet spout.

The work capacity of this machine can be varied by adjustment of the hopper flap. The maximum and minimum rates are given below:
No.2 machine: 52 to 304 kg/h; 1.5 hp
No.3   "   : 254 to 406 kg/h;   2 hp

Source: ITDG, op.cit.

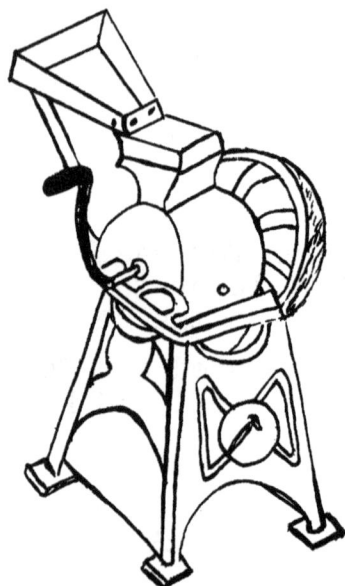

Fig. II.4 (d)

HAND OPERATED GROUNDNUT DECORTICATOR

Manufactured by:

Dandekar Brothers,
Sangli,
Maharashtra,
India

Source: ITDG, op.cit.

FIGURE II.4 (continued)

Different types of groundnut shellers

II.3.5 Storage

The following conditions should be attained, as far as possible, to successfully store groundnuts:

(i)   Tests have indicated that the storage life of groundnuts begins in the field. Therefore, the nuts should have a high initial quality. The ill effects of improper storage, it must be remembered, are cumulative and irreversible.

(ii) The temperature in the storage should be low. Table II.6 shows the relationship between temperature and the length of time the nut retains its edible quality.

(iii) The relative humidity should be between 65-70 per cent. Above 70 per cent, the nuts are likely to grow mould. Below 65 per cent, the nuts lose weight, become brittle and may split during handling.

(iv) The atmosphere in the storage area should be free of odours and well aerated, because nuts readily absorb odours and flavors from the surroundings.

Table II.6:
Relationship between temperature and storage time of groundnuts

Temperature	Time of retaining edible quality	
	Unshelled nuts	Shelled nuts
70°F (21.1°C)	6 months	4 months
47°F (8.33°C)	9 months	6 months
32°F (0°C)	3 years	2 years
25°F (-3.9°C)	7-8 years	5 years
10°F (-12.2°C)	15 years	10 years

CHAPTER III

OIL EXTRACTION FROM GROUNDNUTS AND COPRA

Once the relevant raw materials (copra or groundnut) has been pre-processed, the next stage is oil processing. This chapter considers processing at various scales. However, given the focus of the memorandum on small-scale production, processes relevant to the latter are considered in detail. Medium and large-scale expeller mills, solvent extraction plants and wet-processing methods, which are all suited for medium and large-scale production, are therefore examined only briefly.

The overall process is the same regardless of whether the raw material used is copra or groundnuts. When differences arise due to the type of raw material used, these are explicitly mentioned at each processing stage. The major difference relates to the extraction rates which are more dependent on the raw material than on the process itself.

The three main stages of oil-processing are:

(i)   <u>Pretreatment</u>: namely, the stages prior to the extraction stage such as cleaning, crushing and scorching.

(ii) <u>Extraction</u>: this stage involves the separation of the raw material into oil and residue (cake).

(iii) <u>Post extraction treatments</u>: comprising the packaging of the oil and cake for marketing. Oil refining, common in large-scale production, is not considered in this memorandum

The sequence of the above stages are detailed in a flow chart (see table III.I). The various processes dealt with below follow the same sequence. These stages are described for the following six types of plants: micro-plant, power ghani mill, baby expeller mill, small package expeller mill, medium and large expeller mills, solvent extraction plant and wet-processing.

TABLE III.1

PROCESSING STAGES

1. REMEDIAL DRYING
2. RAW MATERIAL STORAGE
3. CLEANING/INSPECTION
4. CRUSHING/BREAKING UP
5. SCORCHING
6. PRESSING

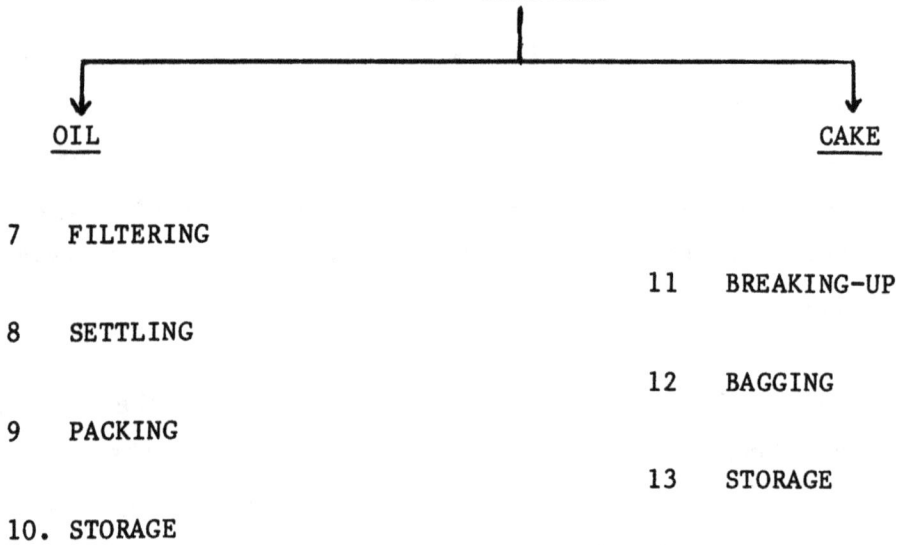

OIL   CAKE

7 FILTERING

                          11 BREAKING-UP

8 SETTLING

                          12 BAGGING

9 PACKING

                          13 STORAGE

10. STORAGE

ANCILLARY SERVICE STAGES

14 WATER

15 PROCESS HEAT

16 ELECTRICITY

17 MAINTENANCE

III.1 : Power ghani mill

The ghani mills originated from India where these indigenous oil crushers have been improved over time. The original animal-powered ghani consists of a wooden mortar and a pestle. The mortar is fixed to the ground while the pestle, attached to one or a pair of bullocks (or buffalos or camel), is rotated in the mortar where the seeds are crushed by the generated friction and pressure. The oil runs through a hole at the bottom of the mortar while the residue - or cake - is scooped out. Depending on the size of the mortar and the type of seeds, an animal-powered ghani can process 5 to 15 kg of seeds at a time. An improved version of the ghani has been developed in India. Known as the Wardha ghani, it is larger and more efficient than the traditional ghani, and can crush charges of seed of up to 15 kg in approximately 1.5 hours or close to 100 kg per day.

An engine-powered ghani is now replacing, to a large extent, the bullock-powered ghani. In this type of ghani, either or both the mortars and pestles are made of cast iron. Power ghanis are often worked in pairs. The crushing capacity of a two-ghanis unit is approximately 500 kg to 600 kg of seeds per day. Some of the technical advantages of the power ghanis, as compared to bullock-powered ghanis include a higher oil extraction rate per unit of raw materials (an increase of approximately 1-2 per cent of the extraction rate), higher output per unit of time, and the use of less space than in the case of bullock-powered ghanis, (at least two power ghanis may be set up in the space needed for a bullock-powered ghani). The quality of the oil produced by the power ghanis is identical to that produced by the bullock-powered ghanis. Power ghanis are now increasingly replacing bullock-powered ghanis in India where the number of the latter has decreased from approximately 400,000 units in the mid-1950s to approximately 230,000 units in the mid-1970s.

Improved power ghanis have an oil extraction efficiency which is fairly close to that of small-scale expellers, and often constitute a viable alternative to the latter, especially in rural areas. While this has been the case in India and in a number of Asian countries, there is no guarantee that ghani mills will meet the same approval in other developing countries. For example, the introduction of ghanis in Tanzania has met with little success. It is therefore important to analyse all the requirements for the successful adoption of ghani mills prior to investing in such units. For example, it is important to investigate whether qualified labour is available in rural areas, whether the repair and maintenance of ghanis can be carried out without much difficulty by the miller or local mechanics and whether power ghanis may be manufactured locally or must be imported.

III.1.1 Pretreatment stages

(a) Remedial drying

Remedial drying may be carried out in the open air or under a covered shed in case of adverse weather conditions. A drying ground of approximately 20 m^2 will suffice (see figure III.1).

(b) Raw material storage

The amount of raw material used requires some provision for storage. To minimise losses due to fungi, insects and the weather, the following type of structures will be needed

- reflective roof
- roof overhang of walls
- controllable ventilation (shutters under the eaves)
- concrete floors for easy cleaning
- pallets or other wooden dunnage for keeping bags off floor; and
- use of fabric insecticides (100 gm of malathron in 5 litres of water, per 100 m^2) on suspect sack material.

The size of the storage area is a function of the amount of stock necessary to ensure continuous production (e.g. stocks for one week or one month). For example, the plant lay-out in figure III.3 shows a storage area of 7.5 m^2.

(c) Cleaning/inspection

Cleaning and inspection are performed manually by removing any large piece of foreign material. Chaff and sand will mostly separate out if the kernels are spread on a dry floor. The floor should preferably be made of concrete; alternatively, any dry packed surface will suffice. The raw materials should be sufficiently clean in order to avoid premature wear and tear of the equipment.

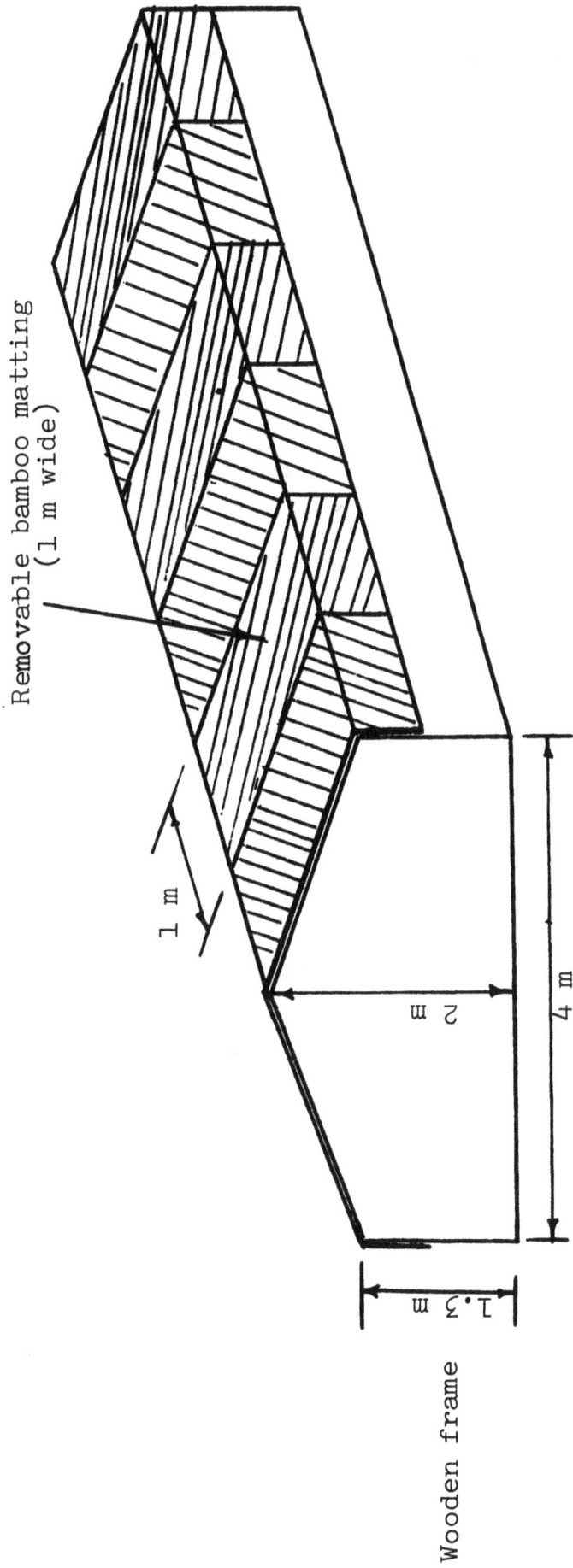

Figure III.1

Trestle and matting arrangements over drying ground

(d) Crushing/breaking-up

Crushing can be done in a small powered hammer mill. These are available from equipment manufacturers in many developing countries. Capacity should be above 100 kg/hour, and reduction should ideally be to pieces less than 0.5 cm square.

It should be noted that pre-crushing is not traditionally used with ghani extraction. However, it is recommended, as its use improves oil yields for ghanis above those assumed in this memorandum.

(e) Scorching

This operation is omitted from the ghani process.

III.1.2 Oil extraction by pressing

The unit considered in this memorandum is a double-ghani mill, powered off a single 3 hp motor. Each ghani takes a charge of about 35 kg which is processed in approximately one hour by the rotary movement of the iron pestle in the bowl. Thus, such a unit may process up to 560 kg of seeds per day. The two ghanis may also be powered by separate engines. In this case, 2 hp motors are needed. The pestles in power ghanis rotate at approximately 10 to 12 revolutions per minute as compared to 3 to 5 revolutions for bullock-powered ghanis.

A large number of designs of powered ghanis have been developed and marketed in India. Although most of these designs are not patented, it would not be practical to provide detailed drawings of these ghanis in this memorandum. Interested readers should write for information to : Appropriate Technology Development Association, P.O. Box 311, Ghandi Bhawan, Mahatma Ghandi Road, Lucknow-226001 (India). This institution may then provide names of ghani manufacturers and/or individuals willing to provide detailed drawings of ghanis for local manufacture.

In order to facilitate oil extraction from copra it is recommended to add 1 percent of "gum accacia" to the raw materials. Thus a charge of 35 kg of copra will require 350 g of gum accacia.

III.1.3 Post-treatment stages

(a) Oil filtering

To remove small impurities, filtering may be done using an ordinary cloth stretched over a frame onto a tank of sufficient capacity (See figure III.2)

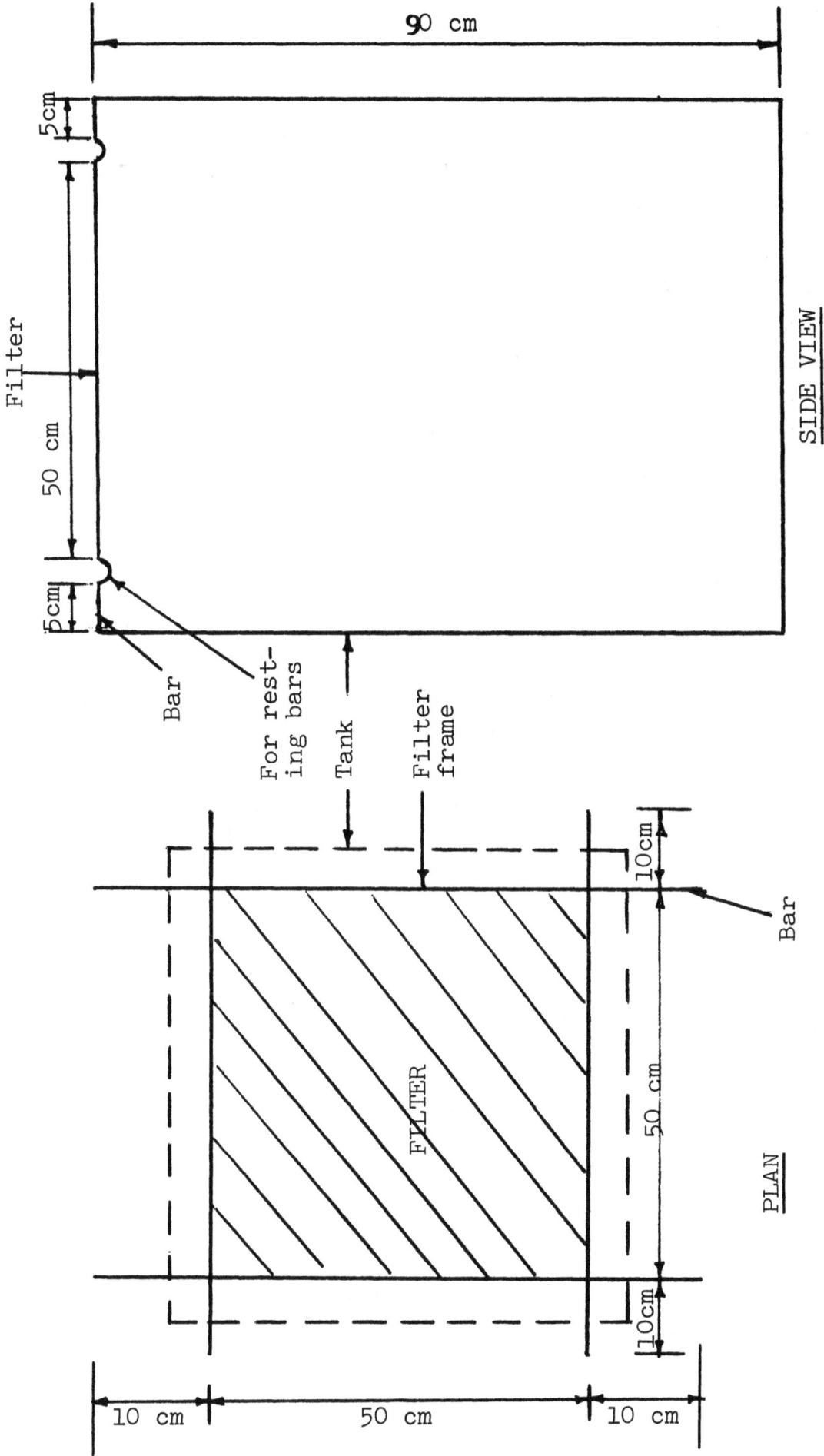

Figure III.2

Filter frame and tank

(b) <u>Settling</u>

The filtered oil should be left in the tank for a few hours in order to allow the settling down of any fine impurities still suspended in the oil.

(c) <u>Packing</u>

The oil is transferred into tins or bottles via a funnel from a tap on the tank. The tap should be attached over the sediment layer.

(d) <u>Oil storage</u> (finished goods)

Storage for several days production will require only an area of a few m^2.

(e) <u>Breaking-up of cake</u>

The cake is removed from the ghani manually. It will probably need little or no further breaking-up.

(f) <u>Bagging</u>

The broken up cake meal can be loaded into bags manually.

(g) <u>Cake storage</u> (finished goods)

A few days' storage space, similar to but separate from the copra storage, will be necessary. Unless a prolonged period before the eventual use of cake is anticipated, insecticides are not necessary. In general, the cake will spoil rapidly after a few days, unless it is properly treated, packaged and stored. Consequently, unless an efficient cake collection system is operating in the country, the produced cake will most probably be used for animal feed within the copra processing area.

III.1.4   Ancillary stage

(a) <u>Water</u>

No special provision necessary

(b) <u>Process heat</u>

It is not essential to use heat in ghani milling. However, it is recommended to warm up the pestle towards the end of the oil extraction process.

(c) <u>Electricity</u>

Required to power hammer mill, ghani and ancillary services.

(d) <u>Maintenance</u>

The mill owner should be able to maintain the power ghanis and undertake standard repairs. If ghanis are not manufactured locally, it should be useful to either import sufficient spare parts or ensure their production by local workshops. This is a particularly important condition for an efficient running of the mill as the international marketing of ghanis (and spare parts) is not yet well developed. Ideally, developing countries wishing to promote the use of ghanis should encourage the local production of the equipment.

III.1.5 Equipment

The following equipment will be required for a plant processing 70 kg of raw material per hour and working an eight-hour day:
- movable scales, capable of weighing up to 40 kg
- 100-200 kg/hour hammer mill, with 6mm square screen (2 hp)
- double-ghani unit, each ghani chargeable with 35 kg. of raw
    material per hour (motor 3 hp)
- 1 filter frame 60 cm x 60 cm with 90 cm bars
- 1 aluminium tank 80 cm x 80 cm x 110 cm

The hammer mill and ghani-unit may need to be imported. Their cost is estimated at US$3,000-4,000.

III.1.6 Daily requirements

The daily requirements for the mill considered above would be:
- raw material: 500-600 kg
- gum accacia: 5-6 kg for copra processing
- diesel fuel: 9 litres
- filter cloth: about two cloths per month
- tins (18 kg.): 18 tins
- electricity: approx 30 kWh.

III.1.7 Labour requirements

Two or three family members will be required for the operation of the mill. One of the workers should have the necessary qualifications for running, maintaining and repairing the ghani unit.

III.1.8 Layout of operations

The power-ghani mill is typically a family enterprise and no formal floor plan is required. However, a plan indicative of good practice is provided in figure III.3.

III.1.9 Daily Output

The daily output of oil and cake are as follows:
- copra processing: 320 kg of oil,
                    213 kg of cake.

- groundnut processing: 220 kg of oil,
                    302 kg of cake.

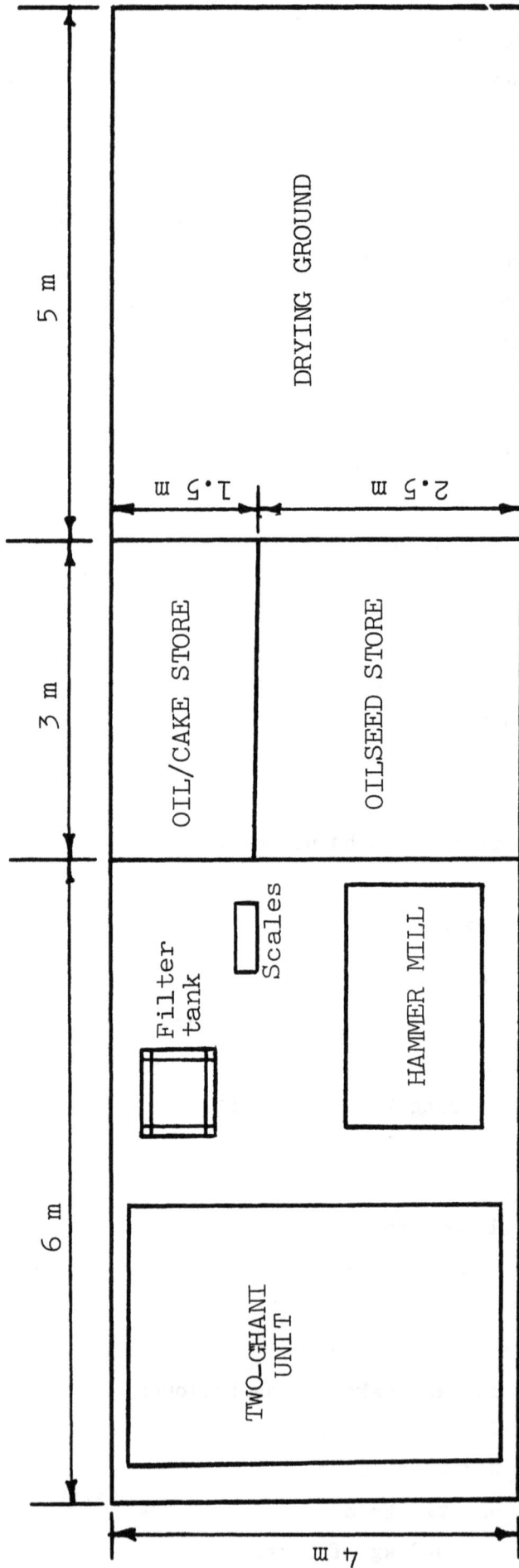

Figure III.3

Plant lay-out for power-ghani mill

III.2 Baby expeller mill

The baby expeller mill considered in this memorandum has a capacity of 45 to 55 kg per hour.   Therefore, by working only one day shift, which is normal for such small plants, the units can process between 350 and 450 kgs of raw material per day.   In a few cases such units do work 2 or 3 shifts per day, and may then process up to one tonne of raw materials.

III.2.1 Pretreatment stages

(a) Remedial drying
Remedial drying is done as in the case of the ghani mill.   The required drying area is 15 m^2.   It should include the necessary structure in case the area must be covered to protect the raw material from the rain.

(b) Raw material storage
As in the ghani mill case.

(c) Cleaning/inspection
As in the ghani mill case.

(d) Crushing/breaking-up
This mill may use a crusher of the swing beater type with a capacity of 100 kg of materials per hour, a 2 hp power requirement, and revolving at 1,400 rpm.   Copra should be reduced to pieces 6 mm square.

(e) Scorching
The oil seeds can be heated in open pans over an enclosed fire place to ensure fuel economy.   The raw materials may be mechanically stirred through, for example, a linkage to the expeller or crusher drive (see figure III.4).

The length of time required is about 20-30 min. and the temperature 60-90°C.   Trial and error will show the best conditions in terms of oil yield and quality.

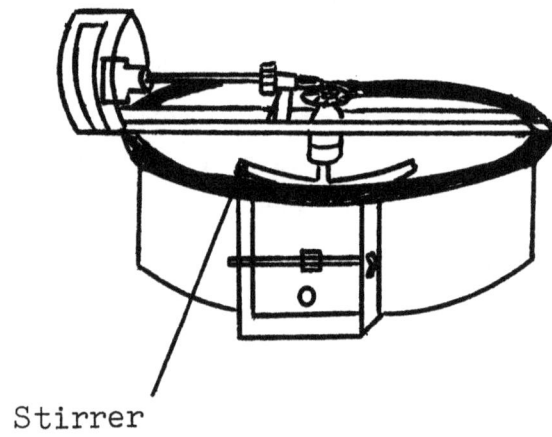

Stirrer

Figure III.4

Single open-type seed scorcher with mechanical
stirrer

Cooking or scorching of oil seed is needed for three reasons: to facilitate oil extraction, to lower or increase the moisture of seed, and to reduce the wear in the screw press. The best temperature and moisture content depend on the extraction system. Copra processed through expellers should have a 2 percent to 3 percent moisture and be at approximately 60°C (higher temperatures for large-scale modern expellers). In the case of groundnuts, the temperatures attained during cooking should not exceed 120°C as otherwise the protein quality may be adversely affected. In general, the required cooking temperature is a function of the cooking time, the type of oil extraction technology, the moisture content of the raw material and the type of seed. All these variables should be considered simultaneously when estimating the cooking temperature.

The scorching of seeds should be carried out with care in order to avoid deteriorating the copra or groundnut kernels. Over-heating or charin reduces the oil extraction rate and yields low quality oil. Since small rural processors may not afford scorching equipment with automatic control of temperature and moisture of the material, they should arrive at the right cooking conditions through learning and experience.

III.2.2 Oil extraction by pressing

Pressing can be achieved by either a single, duo, or duplex expeller. Their drive can be provided by either an electric motor, or via a pulley and v-belt from a separate diesel engine. Small capacity expellers are produced by various specialist manufacturers, notably in Japan and India.

Operations and precise capacities vary from machine to machine. However, the following basic variants occur:

(i) a single expeller capable of producing the assumed oil yield in a single pass. An example of such an expeller is shown in figure III.5 It has the following characteristics:
- capacity: 45-55 kg per hour
- power requirement: 3 hp
- 300 rpm;

(ii) A single expeller of lower pressure but higher first-pass capacity. The scorched seed is fed through once at lower pressure (choke fairly wide open) and the press cake is then

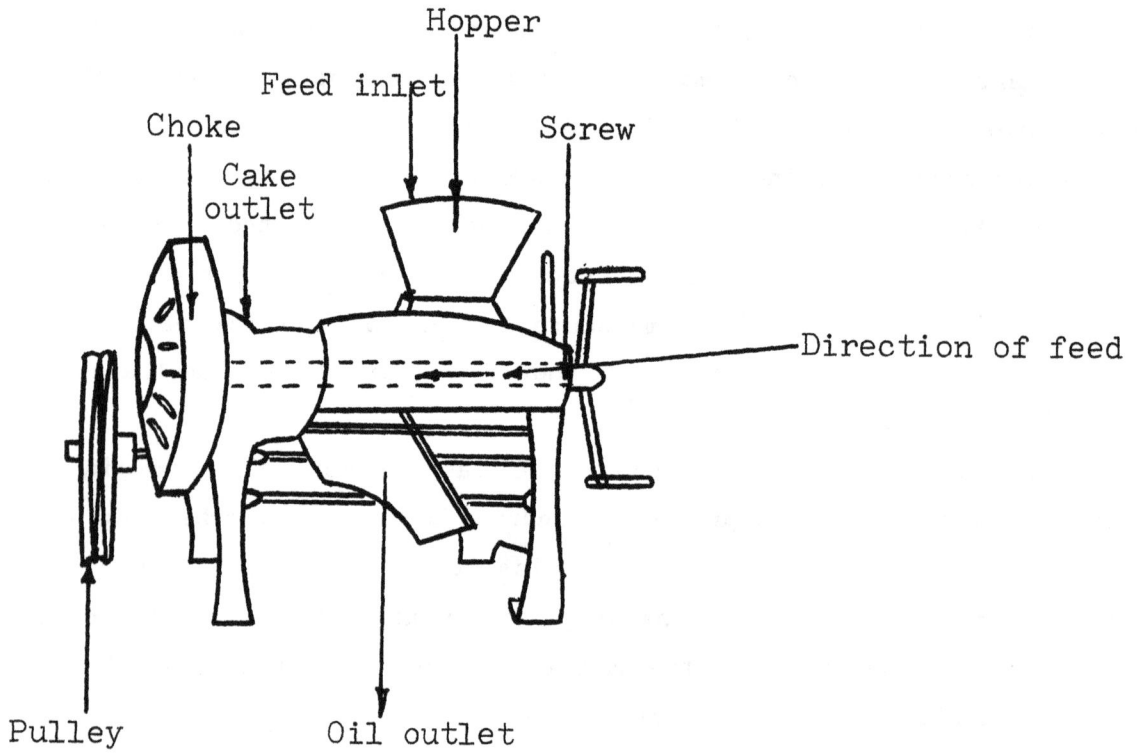

Hopper

Feed inlet

Screw

Choke

Cake
outlet

Direction of feed

Pulley

Oil outlet

Figure III.5

Sketch of a low-pressure expeller   (2 hp)

re-fed through the expeller at a higher pressure (choke narrowed). Most of the oil is extracted in the first pass, but a significant additional amount is yielded by the second pass. Obviously, the expeller must be capable of completing the processing of the daily input in this two-pass manner. Therefore, the capacities of each of the single passes must be higher than the daily input;

(iii) A dual expeller combining a first-pass low pressure expeller and a second pass high pressure expeller. In this case the cake from the first expeller is fed automatically to the second;

(iv) A duplex expeller employing screws of varying pitch and differently shaped barrel areas. This arrangement increases the pressure as the material is forced through the barrel. Duplex units achieve higher yields and faster throughput in a single operation. They are recommended if the right capacity machine is available.

When using a single expeller, the decision on whether to pass the oil seed once or twice should be based on economic considerations. A second pressing raises the oil extraction rate, and therefore yields additional revenues, but also increases processing costs. Thus, a second pressing will be justified in the only case where the increase in revenues is at least equal to the increase in cost. In general, given the relatively low extraction rate of small expellers, small rural mills in developing countries find it profitable to press the raw materials twice, the first time at low pressures - and therefore high hourly output - and the second time at higher pressures through an adjustment of the choke.

The extraction of groundnut oil in small expellers is a highly skilled job as it is necessary to add groundnut shells to the kernels in order to prevent the forming of peanut butter. The adding of shells is necessary because groundnut kernels, unlike copra, have little fibre. The need for skilled labour should therefore be taken into consideration when choosing among various types of expellers.

When ordering the equipment, the mill operator should specify the type of raw materials to be processed as the spacing between the bars (which are part of the expeller barrel) is a function of the type of oil seed. For example, the spacing of the bars in the feed and centre sections of the expeller are approximately three times wider for copra than for cotton seed.

Expellers require periodic maintenance and repairs, the principal wearing pieces being the liner bars, the worms and the distance pieces. The periodicity of maintenance and repairs is a function of the rate at which the above piece of equipment is worn out by the abrasive action of the raw materials and that of foreign matter (e.g. sand, pieces of iron). The abrasive action of foreign matter is particularly harmful and can considerably shorten the life of various parts of expellers.

Line bars suffer less than other parts of the expeller, but cannot be repaired: they must be replaced by new bars. On the other hand, worms and distance pieces can be welded and re-used. In general, worms may work eight to ten weeks before repairs are needed, depending on the type of expeller. However, foreign matter may shorten the above period to two weeks or less. It is therefore essential that the raw materials be as clean as feasible under the prevailing processing conditions.

Mill operators should learn to carry out basic repair and maintenance of expellers or be able to secure the services of a nearby specialised mechanic. They should also have available a stock of parts if the latter are not manufactured locally and may not be obtained as soon as they are needed. It is essential that repair and maintenance of expellers can be carried out quickly if one is to avoid high processing costs (i.e. through an increase of depreciation costs resulting from a lower utilisation rate of the expeller).

The choice of expeller is a difficult one to make as the mill owner should consider both the current scale of production as well as an eventual expansion of the mill. Although no precise advice may be given without knowing the exact circumstances under which a mill is being established, a general rule for small rural mills is to start with a relatively cheap expeller which may process, in one or two passes, the daily input of raw materials. As suggested earlier, the decision regarding a second pass of the cake should be based on a comparison of additional revenues and costs. The lay-out of the mill should, however, be such as to allow the use of a second expeller if an expansion of capacity is being considered. The prime mover of the expeller should also be suitable for the second expeller. The latter may have a higher capacity than the first one in order to ensure greater flexibility in the mill operation.

III.2.3   Post-treatment stages

(a)   Filtering (oil):As in the case of the ghani mill. Alternatively, and whenever it is economically feasible, the mill may use a small chamber filter press with ten plates of 18 x 18 cm, with attached oil pump. The pump makes 170 rpm and has a power requirement of 0.5 hp. Such a press may process up to 50 litres per hour, and should therefore be sufficient for the filtering of the daily output of oil. Filtration takes place through paper and filter cloth.

(b)   Settling : As in the case of the ghani mill.

(c)   Packing : As in the case of the ghani mill.

(d)   Oil storage.: As in the case of the ghani mill.

(e)   Breaking-up of cake, bagging and cake storage: As in the case of the ghani mill.

III.2.4   Ancillary stages

(a) Water

   No special provision necessary.

(b) Process heat

   The amount of process heat depends on the adopted scorching technique.

(c) Electricity

   Requirements for expeller drive.

(d) Maintenance

   The mill owner should be familiar with the maintenance and repair of expellers and should have sufficient spare parts on order to avoid long shut-downs of the mill. The type of repairs which may be needed have already been described earlier in this section.

III.2.5 Equipment

   The following equipment will be needed for a plant processing 400 kg of raw material per eight-hour working day:

- movable scales, capable of weighing up to 80 kg

- one seed crusher of the swing-beater type, 2 hp, 1400 rpm

- one seed scorching pan, 100 cm diameter x 15 cm deep, with a mechanically driven stirrer.

- one expeller unit, power driven,with a processing capacity of 50 kg of raw material per hour (3 hp). It should produce cake with an oil content of 7.5 to 8 per cent

- one filter frame, 60 x 60 cm, with 90 cm bars
- one aluminium tank 80 x 80 x 110 cm.

    The total cost of the equipment which may need to be imported is estimated at US$ 12,000.

III.2.6 Daily requirements

    The daily requirements for a plant similar to the one considered above will be:-
- raw materials: 400 kg
- diesel fuel: 9 litres
- electricity (if no diesel fuel is used) : 30 kWh
- wood or other local fuel: 30 kg
- tins (18 kg): 14
- filter cloths: three every two months.

III.2.7 Labour requirements

    Two or three family members should suffice for the running of the unit. The mill owner should have the necessary skills for the running of the equipment, as well as for maintaining and repairing the latter.

III.2.8 Layout of operations

    This scale is typically a family enterprise. Therefore the floor plan in figure III.6 is only indicative of good practice.

III.2.9 Daily output

    The daily output of oil and cake are as follows:
- copra processing: 245 kg of oil,
                  135 kg of cake.
- groundnut processing: 165 kg of oil,
                  215 kg of cake.

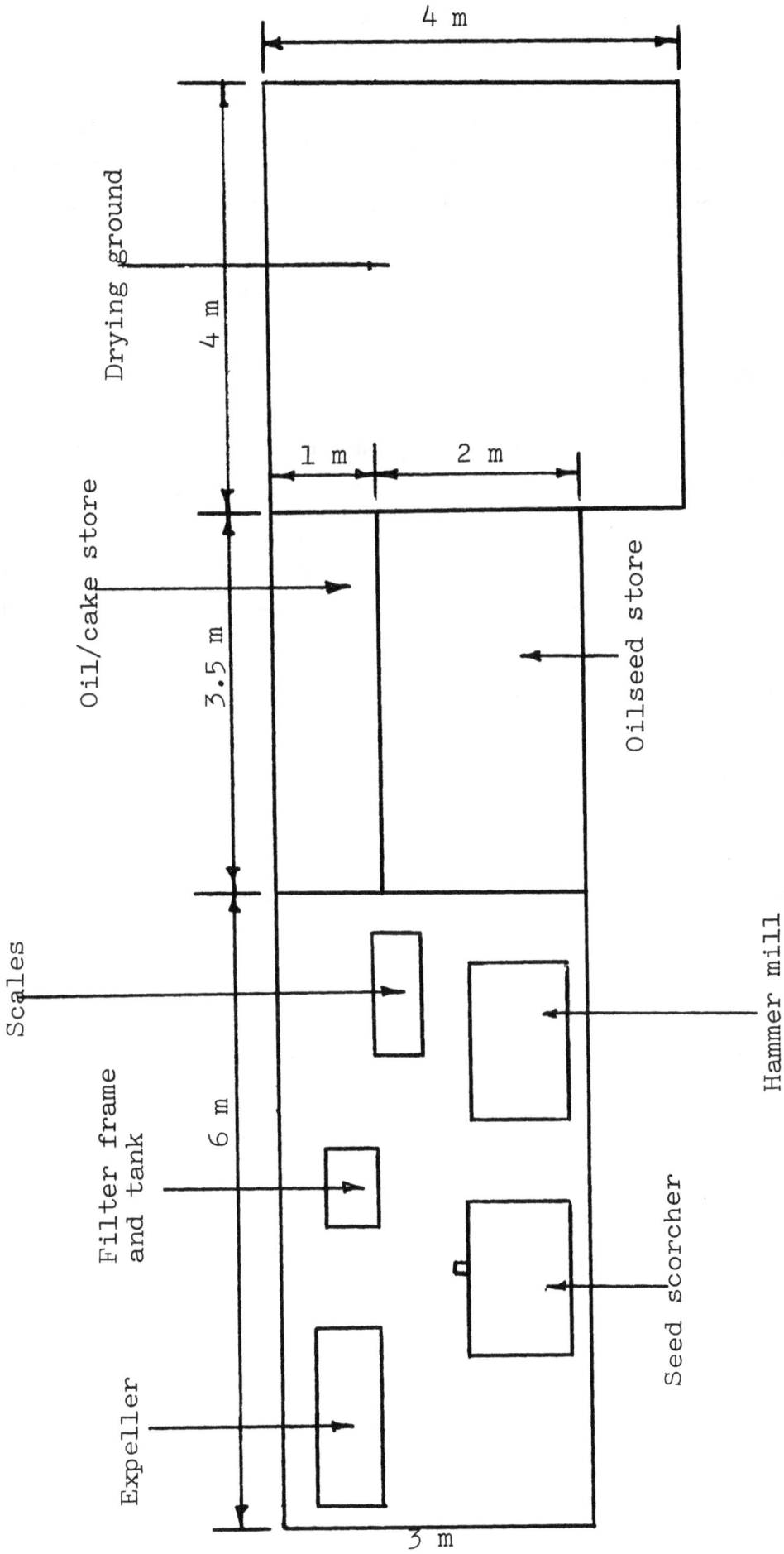

Figure III.6

Plant lay-out for a baby expeller mill

III.3   Small package expeller mill

The capacity of this mill is the double of that of the baby expeller mill (800 kg/day).   Therefore, certain modifications to the process described above are required for this type of mill.   The capacity of this mill may be increased to 2.4 tonnes per day on a three-shift basis. This will require an extension of the storage areas for raw materials and output.

III.3.1 Pretreatment stages
(a) Remedial drying

The method described in the previous plants may still be adequate.   It may also be possible to employ a simple form of in-sack artificial drying by building a platform above the seed scorchers (see figure III.7).

(b) Raw material storage :As in previous plants.   Sufficient space should be allowed for the storage of one week's input of raw materials (e.g. 4 tonnes).

(c) Cleaning/inspection : As in previous plants.

(d) Crushing/breaking-up

As in the previous plants.   The crusher capacity should be scaled up to allow a fast completion of the operation, thus freeing labour for other operations.

(e) Scorching

As in the case of the baby expeller.   Two scorcher units are used side-by-side at this scale.

III.3.2. Oil extraction by pressing

A motor-driven expeller capable of processing 120-140 kg/hour of raw material is required.   A duplex unit is most suitable for this purpose. If a single expeller were to be used, it should have a power requirement of 5 hp and run at 350 rpm.

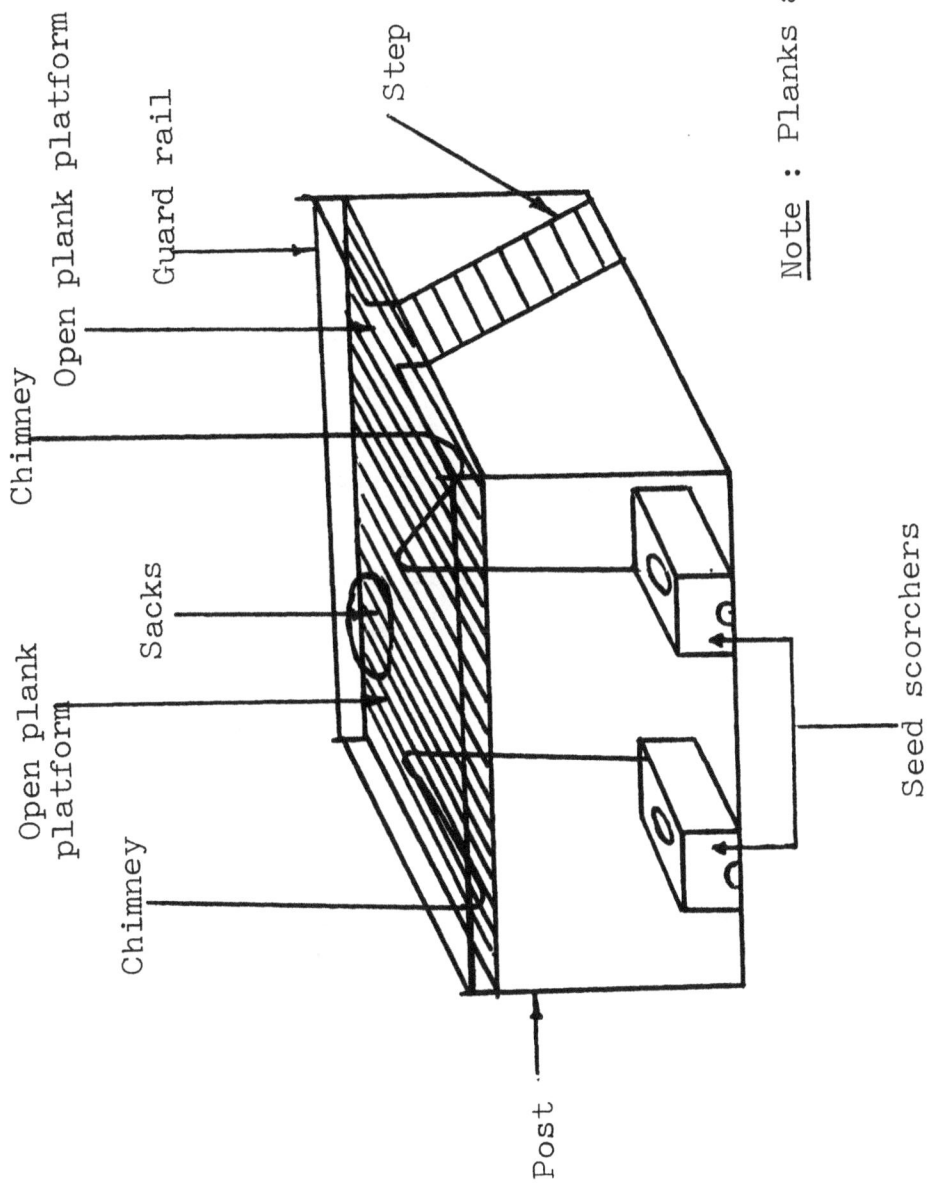

Note : Planks are 4 cm wide and 15 cm apart

Figure III.7
Copra drying platform over scorchers

III.3.3. Post-treatment stages

(a) Filtering (oil)

The oil is passed through the expeller screen into a small holding tank from where it is pumped mechanically through a small chamber filter press. A filter press twice the capacity of that described in the case of the baby expeller should be adequate. The residue which accumulates on the cloth undergoes a second passage through the expeller.

(b) Settling

After the filtering, it is best to leave the filtered oil to settle for a couple of days. Two separate tanks are used for this purpose. Provision should be made for periodic draining of sediment through taps at the bottom of the tanks. The latter should be inclinable to facilitate this operation (see figure III.8).

(c) Packing

At this scale, the same operation as used in plant 2 is adequate. However, a small foot-operated or motorised pump would facilitate this operation.

(d) Oil storage

The plant is unlikely to store more than a few days production as distribution is likely to be largely local. However, some of the oil may be marketed more widely, and sufficiently large storage capacity may be required.

(e) Cake breaking-up: As in the case of the baby expeller.

(f) Bagging: As in the case of the baby expeller.

(g) Cake storage: As in the case of the baby expeller.

III.3.4. Ancillary stages

(a) Water

No special provision necessary unless the adopted expeller is water-cooled.

(b) Electricity

Electricity is required to operate the equipment.

(c) Maintenance

As in previous plants. At least one operator should be skilled in small mechanical repairs to reduce expenses.

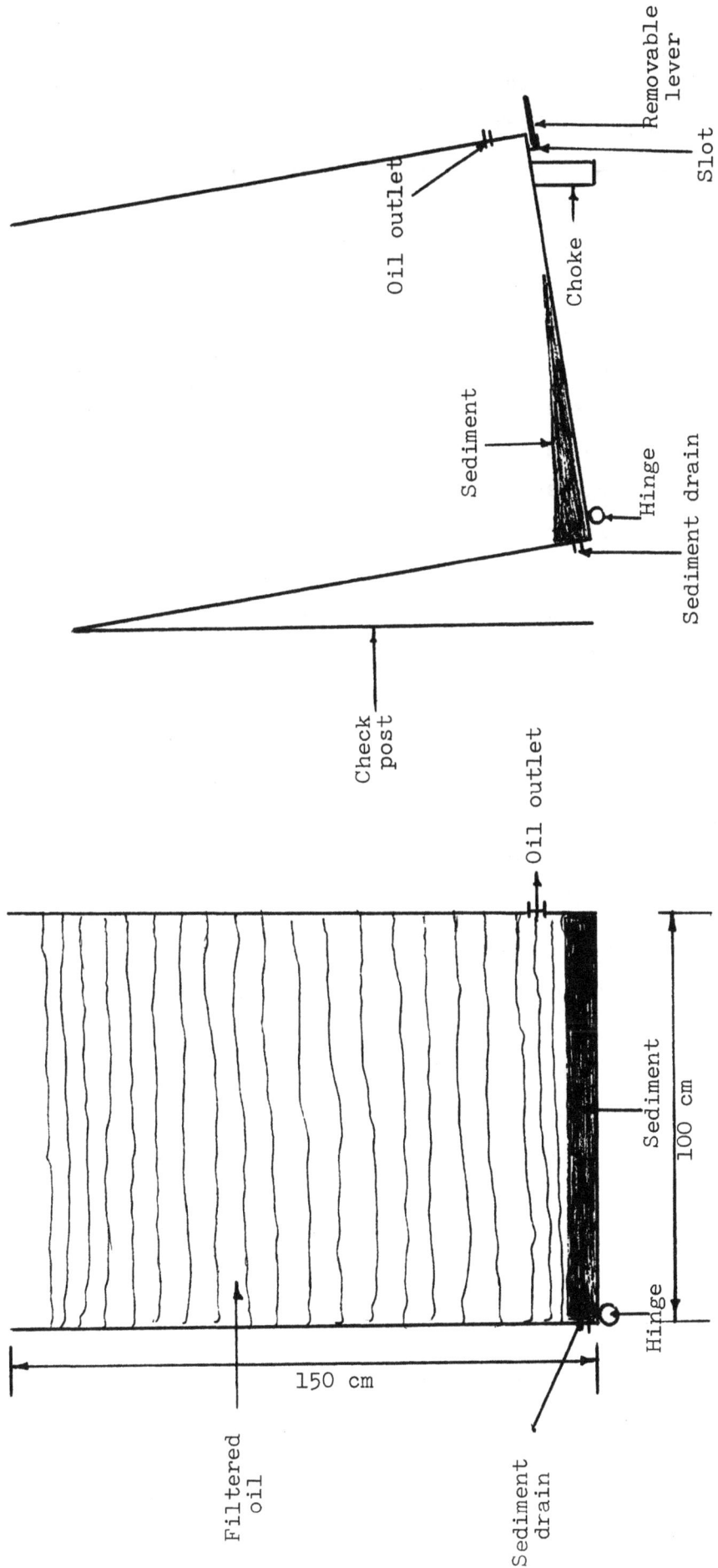

Figure III.8

Inclinable settling tank for sedimentation drainage

III.3.5. Equipment

The following equipment will be needed for a plant processing 800 kg of raw material per eight-hour working day:

- movable scales, capable of weighing up to 80 kg
- drying platform over seed scorchers, if desired
- one 200-400 kg/hr crusher mill with a 6 mm square screen (4 hp)
- two seed scorching pans, 100 cm diameter x 15 cm deep with mechanically driven stirrer (drive run off crusher mill power by v-belt and pulley)
- one duplex expeller unit, with an input capacity of 100 kg/hour(6 hp). The oil content in the cake should be 5-6 per cent. Alternatively, a 5 hp expeller may also be adequate
- one oil holding tank 75 x 75 x 90 cm
- one champer filter press with 12 plates 25 x 25 cm to filter
         up to 100 or more litres/hour (this allows room for expansion of
         expeller capacity), with attached oil pump
- two aluminium settling tanks, 100 x 100 x 120 cm (capable of
         holding together up to two day's oil output), with oil outlets and
         sediment drainage
- one small pump with plastic hose for pumping filtered oil into
         tins.

The total cost of equipment which may need to be imported is estimated at US$32,000.

III.3.6. Daily requirements

The daily requirements for a plant like the one considered above would be:

- raw materials: 800 kg
- electricity: 75 kWh
- wood or other local fuel: 60 kg
- filter cloths: 5 per 2 months
- tins (18 kg): 28.

III.3.7. Labour requirements

Three workers. One should be skilled in simple maintenance and repair tasks, especially the repair of the worms and bars.

III.3.8. Layout of operations

The floor plan in figure III.9 is indicative of good practice. If desired, the layout may be spaced out to allow for expansion of capacity.

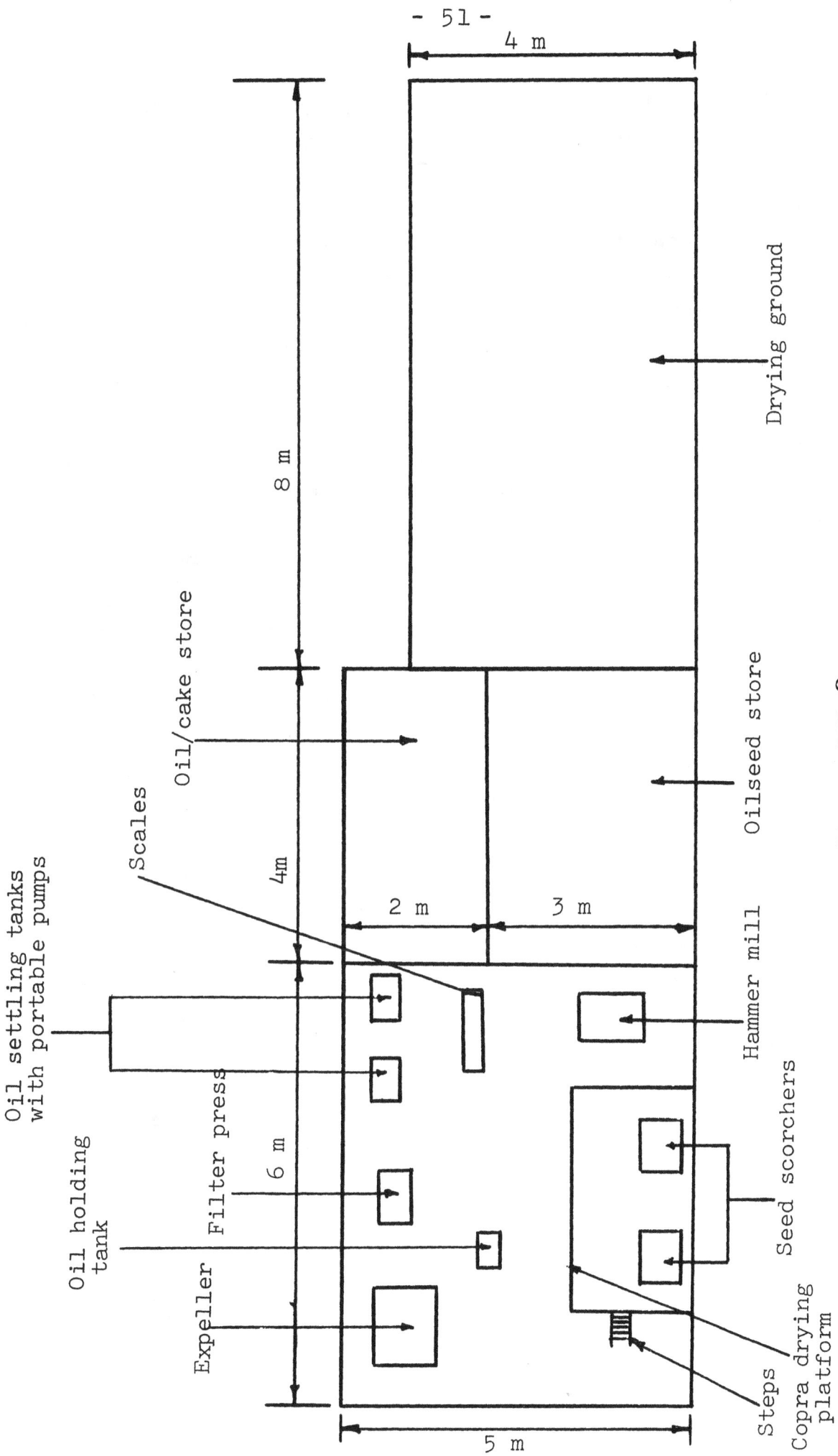

Figure III.9

Plant lay-out for a small package expeller mill

III.4 Medium and large expeller mills, solvent extraction
plants and wet-processing plants

Both medium and large expeller mills follow the same stages of oil processing as the three plants considered above. As is to be expected, their scales involve the use of larger and more sophisticated equipment. Technical details on these mills are not provided in this memorandum since the latter is to focus essentially on small-scale oil extraction units.

III.4.1. Medium expeller mills

The lowest economical capacity for a medium-sized oil mill, working one day shift, is three-and-a-half tonnes, i.e. 1,000 tonnes per year. However, it is fairly normal for this type of plant to operate on three shifts per day and to process eight to ten tonnes per day, or 2,400 to 3,000 tonnes per year. The upper limit is about 20 tonnes per day, i.e. 6,000 tonnes per year.

Some pieces of equipment not used at small scales of production are needed at these larger scales. They include seed cleaners, a separate screening tank, an air compressor for the filter press and conveyors for the mechanical transport of various materials including the expeller cakes. A list of equipment needed for a 20 tonnes per day coconut oil extraction plant is shown in table III.2. This list is taken from the following UNIDO publication:

UNIDO: Coconut Processing Technology Information Documents,

Part 2 of 7, Coconut oil extraction (Vienna, 1980),

limited distribution.

The above publication focuses on medium to large-scale oil processing plants, providing detailed technical information on nine different oil extraction units ranging from a full-press mechanical extraction plant of 4 tonnes copra/day to a full solvent extraction plant of 150 tonnes/day. A 250 tonnes copra/day wet-coconut processing plant is also described in this publication.

Table III.2

List of equipment for a 20 Tonnes/day
copra processing plant[1]

Number of
pieces

2 — Platform scales, beam-type, capacity 500 kg, 0.1 kg accuracy for weighing copra in sacks.

1 — Copra bin, 20 metric tons capacity in three compartments, mild steel construction, with bottom conveyor.

1 — Copra pre-crusher, horizontal, low-speed hammer mill; capacity 1 ton/hr; with hopper, 10 hp motor drive.

2 — Copra grinder, high-speed peg mill, horizontal shaft, with 100 cu ft hopper, screen feeder, each with 15 hp motor drive; capacity 1 ton/hr each. Used alternately.

1 — Roller mill, 3-high, with 3 sets of 10" x 3' rollers, 100 cu ft hopper and 7.5 hp motor drive; capacity 1 ton/hr.

2 — Expellers, single worm, 4-high croker-conditioner, vertical feeder, distribution conveyor, overflow bin, return conveyor, each with 25 hp motor drive; capacity 500 kg/hr each.

1 — Screening and settling tank, 20-mesh stationary screen on top of the tank, 100 cu ft. tank with conical bottom. foots withdrawal bottom valve; mild steel construction.

1 — Filtering pump, centrifugal, 60 psi, capacity 15 gpm, cast-iron casing, cast-iron impeller, graphited asbestos packing, open-impeller type, with 2 hp motor drive.

1 — Plate-and-Frame Filter Press, flushed-plate type, open delivery, 24" x 24" with 30 plates and 29 frames, with oil trough, and cake pan.

1 — Crude oil storage tank, 100 tons capacity, cylindrical with conical top, depth gauge, mild steel construction.

1 — Cake grinder, made of ribbon type screw conveyor with rod breakers, 3 hp motor drive.

1 — Meal bin, cylindrical with conical bottom discharge gate, capacity '5 metric tons of cake, mild steel material.

1 — Platform scale for weighing cake in bags, capacity 500 kg, beam-type, 0.1 kg accuracy.

Table III.2 (Cont'd)

1 -   Platform scale for weighing coconut oil in 55-gal drums beam-type; capacity 500 kg.

1 -   Set of conveyors: inclined screw conveyor from bodega to copra bin, transfer screw conveyor from bin to crusher, rotor-lift from crusher to grinder, rotor-lift from grinder to roller, horizontal screw conveyor and rotor-lift from roller to cooker.

1 -   Oil transfer pump for pumping oil from filter to storage tank, centrifugal type, 60 psi, CI casing and impeller; capacity 15 gpm; with 2 hp motor drive.

1 -   Oil transfer pump for pumping oil out of storage tank, capacity 30 gpm, centrifugal, CI material, with 3 hp motor.

1 -   Set of drum-closing tools to seal 55-gal steel drums.

Source: UNIDO (1980), op.cit.

The daily requirements, including labour, for a medium expeller mill are given in Chapter IV.

III.4.2. Large expeller mills

Large expeller mills are larger versions of the medium expeller mills considered above. The lower economic level for such plants is 22 tonnes of raw material processed per 24 hours. The use of three duo super expellers could raise the production level to nearly 100 tonnes of raw material processed per 24 hours (three shifts)

The equipment for a plant operating at this scale would include large expellers, the equipment mentioned in connection with medium expellers (suitably scaled up) and new features such as automatic scales, a separate cooker dryer and oil cooling of the expeller

Daily requirements, including labour, for a plant at this scale are given in Chapter IV.

III.4.3. Solvent extraction plants

Even the most perfect expellers leave at least six percent of oil in the expeller cake. The above rate is normally exceeded in most expellers. It is possible to recover these losses using a solvent extraction plant. Such a plant can reduce the residual oil in cake to less than 1 percent although it is good practice to maintain a 2 percent oil in cake. However, a major drawback of this process, especially in view of the bias of this memorandum towards small-scale production, is that it is by nature suited to large scale extraction. The range of production levels employed in solvent extraction plants is between 50 to 200 tonnes per day. Other drawbacks include the high investment costs, the need for highly skilled labour, low employment generation and danger of explosions if the plant is not kept in perfect condition. While solvent extraction plants may not constitute a viable rural industry, a number of these plants (depending on the national volume of oil seeds processed each year) may be profitably established for the processing of both oil seeds and the cake produced by small oil extraction units. This will, however, require a good cake collection system and a sufficient supply of oil seeds in order to maintain the solvent extraction plants running at sufficiently high capacity utilisation rates.

There are two main groups of solvent extraction plants:

- batch extraction

- continuous extraction.

Common to all extraction plants is their minimal need of operators for supervision. However, the operators that are needed require long and careful training.

III.4.4. Wet-processing (for copra only)

The technique of wet-processing involves the mechanical conversion of fresh coconut meat into a milky water-based emulsion which is separated into oil and protein by centrifugation. If a solid protein product is desired, a further drying stage is necessary.

The motivation for wet-processing is the recovery of protein contained in copra. The latter tends to be denatured by the conventional process of copra milling. Furthermore, the traditional copra milling process entails losses of oil from spoilage during the copra drying stage, from insect or rodent attack during the copra storage stage, or from incomplete extraction of oil by mechanical as opposed to chemical means of oil extraction. Obviously, these losses can be minimised if processing is undertaken with care.

Wet-processing has many drawbacks. Its major drawback, from the point of view of this memorandum, is the fact that they are not economically feasible at small scales of production. In addition, wet processes entail a loss of oil which is rarely lower than 10 per cent. Another drawback is that the process involves the use of fresh coconut meat. This normally implies higher transport costs than those involved in copra milling.

CHAPTER 4

ECONOMIC ASPECTS OF OIL PROCESSING

This chapter is intended for practising or potential millers wishing to compare alternative oil extraction technologies suited for small-scale production. Thus, the following two sections provide a general methodological framework which may be applied by the reader to local conditions and circumstances, taking into account local demand, factor prices (e.g. wages, unit prices of various inputs), local interest rates, etc. An illustrative example of this methodological framework is provided in section IV.3 in relation to a two-ghani power mill. Section IV.4 provides an economic evaluation of the processes described in Chapter III, on the basis of a given set of factor prices and interest rates. The above section may be of interest to public planners who must decide on the type of plants which should be promoted in the country. Finally, section IV.5 provides a sensitivity analysis with a view to showing how the economic feasibility of various oil extraction technologies may differ with varying wages and interest rates.

IV.1 Methodological framework for the estimation of production costs and revenues

The methodological framework considered below is of a general nature and may be used to estimate production costs of groundnut as well as coconut oil.

Step 1

(i) To determine the daily and yearly quantities of raw materials (groundnuts or copra) to be processed, taking into consideration the local availability of materials and intended capital investments.

(ii) To determine the number of shifts worked per day and the number of days worked per year.

(iii) To calculate the output, using the extraction rates associated

with the adopted process (see table IV.1), and the input of raw
materials:

- oil output = extraction yield of oil x M tonnes/year of raw
  materials.

- cake output = extraction yield of cake x M tonnes/year of raw
  materials.

The extraction rates in table IV.1 should be considered as average
rates obtained under fairly good processing conditions. These rates were
obtained from field studies reported in a number of publications including
Thampan (1975), FAO (1971), Thieme (1968), UNIDO (1977) and UNIDO (1979).
Under inadequate processing conditions, the oil content of cakes may be as
high as 20 percent to 30 percent in the case of power ghanis, and as high as
12 percent in the case of medium/large expellers.

Step 2

To determine fixed investment cost. This step consists of determining
the costs of the following items:

L = Cost of land

B = Cost of Buildings

D = Cost of drying grounds

E = Equipment cost (local and imported)

S = Cost of initial spare parts (equal to 5 percent of initial imported
equipment). Therefore Fixed Investment Cost (FI) = L + B + D + E + S.

Step 3   Working capital determination

The working capital required is dependent on the adopted levels
of stocks of finished goods and raw materials. It may be estimated on the
basis of the following formula:

Working capital = 1.1 x A x M/day x C

where:

A = The sum of the number of days during which raw materials
input and the finished goods are kept in storage.

M/day = Amount of raw materials input (in tonnes)

C = cost of 1 tonne of raw materials input.

Table IV.1

Oil extraction rates attained by selected technologies

Percentage/ Processes	Extracted oil: percentage of total raw materials input	Cake: percentage of total materials input	Percentage of oil in cake	Technical efficiency as a percentage of total oil content[1]
(a) Coconut (copra)				
Small-powered expellers (Plants 2 and 3)	61.3	33.0	8.0	95.8
Power ghani	57.0	33.0	18.4	89.3
Medium/large expellers (Plants 4 and 5)	62.0	33.0	6.0	96.9
(b) Groundnuts				
Small-powered expellers (Plants 2 and 3)	41.0	54.0	7.5	91.1
Power ghani	38.9	54.1	11.0	86.4
Medium/large expellers (Plants 4 and 5)	42.4	52.6	5.0	94.2

---

[1] Oil contents of 64 per cent for copra and 45 per cent for groundnuts are assumed for the purpose of the economic analysis of alternative oil extraction technologies. These percentages refer to moisture free raw materials.

Step 4

To estimate total investment cost as the sum of investment costs and working capital.

Step 5

To determine the fixed investment annual cost for every investment component. This cost is a function of the interest rate assumed. Let this be I per cent per year. Given the value of I and knowing the useful life of the piece of equipment, the annual fixed investment cost of the latter can be calculated in the following manner: For interest I per cent p.a and a useful life U, to obtain the corresponding factor F from a discount table (see appendix). For example, given a value of I of 10 per cent and a useful life of 10 years, we find F = 6.145. Let the investment cost of the component be Z: then the annual fixed investment cost is equal to Z/F. In this manner, one may calculate this cost for each investment item including building, drying grounds and equipment. Since land has an infinite life, the annual cost may be assumed to be equal to the annual rental rate. The annual cost of spares and maintenance - which may be assumed to be equal to 7.5 percent of total equipment costs - should be added to the other annual fixed costs in order to obtain the total annual fixed costs.

Step 6

To estimate the working capital annual cost as the annual interest paid on the amount of working capital.

Step 7

To estimate total fixed annual costs as the sum of annual fixed costs (step 5) and interest paid on working capital (step 6).

Step 8

Next, the sum of the annual variable costs is calculated. These include the annual costs of: raw material((M tonnes/year) x (cost of 1 tonne)); water; electricity; diesel; wood or other local fuel; filter cloth requirements; tins or drums; and labour (number required x yearly wage).

Step 9

To estimate the total annual costs as the sum of annual fixed costs and annual variable costs.

Step 10

To estimate total annual revenues as the sum of two components:

- revenues from the sale of oil
- revenues from the sale of cake

The current market prices of oil and cake may be used to estimate total revenues.

Step 11

To estimate the annual gross profits by subtracting total revenues from total costs.

This general methodology is illustrated by an example in section IV.3.

IV.2 Assumptions made in economic evaluation of alternative oil extraction technologies

An economic comparison and evaluation of various processes implies the use of certain assumptions. This section deals with the economic assumptions used in this memorandum. Care has been taken to ensure that the assumptions involved reflect, to a large extent, conditions prevailing in developing countries that produce coconut or groundnut oil. The assumptions are applied to all the processes considered, except solvent extraction (of groundnuts and coconuts) and wet-coconut-processing. These two oil extraction technologies are not evaluated for the following reasons.

Firstly, full solvent extraction of oil from high-oil-bearing seeds (e.g. copra and, to some extent, groundnuts) is rarely practised in oil extraction mills for cost reasons. Thus, the oil seeds are first processed in expellers where 85 percent of the oil is extracted. The oil remaining in the cake is then processed in the plant solvent extraction unit. The decision to invest in a solvent extraction plant is therefore, to a large extent, independent of that to invest in expellers (small or large) or ghanis, inasmuch as the former plant is generally complementary to the latter ones. Rather, the decision to invest in a solvent extraction plant should be based on a comparison of costs and revenues associated with the following two alternative choices:

- oil extraction by expellers or ghanis only;
- oil extraction by expellers or ghanis, followed by solvent extraction.

Let us assume that the choice is between the use of only an expeller with a 95 per cent oil recovery efficiency or the use of both an expeller with an 85 percent oil recovery efficiency and a solvent extraction unit (recovery of an additional 14 percent of oil from the cake). We then have the following costs and revenues per tonne of raw materials processed:

C_1 = processing cost of raw materials for 95 per cent oil recovery by expeller;

C_2 = processing cost of raw materials for 85 per cent oil recovery by expeller;

C_3 = cost for treating cake (15 per cent oil content) from expeller in solvent extraction unit (1 per cent oil remaining in cake after extraction);

R_1 = additional revenues from the sale of additonal oil (4 per cent) recovered by solvent extraction, when compared to 95 per cent extraction by expeller;

R_2 = revenue from the sale of cake if only expeller is used;

R_3 = revenue from the sale of cake if both an expeller and a solvent extraction unit are used. Generally, $R_2 > R_3$ due to the higher oil content of cake from expeller.

A solvent extraction unit will then be profitable if the following relationship holds:

$$R_1 + (C_1 - C_2) - C_3 - (R_2 - R_3) > 0$$

Secondly, economic evaluation of wet-coconut-processing is not carried out because this technology is not yet well established in developing countries and it was not therefore possible to obtain reliable data on raw materials costs (fresh coconuts), plant costs, labour input, productivity, revenue from the sale of flour, etc. This does not, however, mean that wet-coconut-processing is not profitable. Interested readers may obtain information on this technology from available publications (see Bibliography) and equipment suppliers.

The various types of plants described in Chapter III are assigned the following numbers for easier future reference:

Plant No. 1 : Power ghani mill

Plant No. 2 : Baby expeller mill

Plant No. 3 : Small package expeller mill

Plant No. 4 : Medium and large expeller mill

Plant No. 5 : Large expeller mill

IV.2.1. Fixed investment cost factors

The assumptions made in this memorandum relate to the following:

(i)   Land: The cost has been assumed to be US$5/m^2. The useful life is infinite.

(ii) Buildings: The cost has been asumed to be US$80/m^2 in rural areas (Plants 1,2,3) and US$100/m^2 in urban areas (Plants 4 and 5). The useful life for all buildings is estimated at 30 years.

(iii) Drying grounds: The cost has been assumed to be US$10/m^2. Their useful life is estimated at six years.

(iv) Initial cost of spares for imported equipment: estimated at 5 per cent of imported equipment cost.

The imported equipment costs are provided below for the 5 plants:

Plant No.	1	2	3	4	5
Cost (US$)	3,500	12,000	32,000	500,000	1,600,000
Useful life(years)	10	10	10	15	15

The above costs, to which must be added the local equipment costs, apply equally to copra and groundnut processing.

IV.2.2. Interest rates and discount factors[1]

The interest rate has been assumed to be 10 per cent p.a. For this rate and assumed lives of buildings and equipment, one obtains the following discount factors, F:

- Buildings: F = 9.427
- Drying grounds: F = 4.355
- Equipment: For plants 1,2,3,: F = 6.145.
            For plants 4 and 5: F = 7.606

Since land has an infinite life, the cost of land may be assumed at 100 of its value (i.e. interest paid on purchase price).

---

[1] The discount factor, F is used to calculate the present value of building and equipment since the latter may have diferent useful lives. A table in Appendix I provides the value of F for different interest rates and numbers of years.

IV.2.3. Raw material prices

Evidence available from major raw material producing countries (e.g. the Philippines and Indonesia) does not indicate an obvious pattern of consistency of local prices in separate countries with one another or with the world market price. Available price data[1] offer evidence that local prices of copra and groundnuts, in developing countries, are significantly below prices in Europe or North America. Thus the following price structure has been assumed:

- copra: US$247 per metric tonne (33 percent below world market prices). This wide spread between local and world market prices is indicative of the existing complex marketing chain, including the involvement of a number of intermediaries between producers and importers.

- Groundnuts: US$470 per metric tonne (25 percent lower than world market prices).

IV.2.4. Annual cost of various inputs used in oil extraction and packaging

The following annual costs - expressed in US$ - are assumed for the purpose of evaluation:

- water: US$1 per m^3
- electricity: US$0.09 per kWh
- diesel fuel: US$0.40 per litre
- wood (local fuel): US$20 per metric tonne
- filter cloths:

    for plants 1 and 2: US$1.50 per cloth
    for plant 3:   US$5 per cloth
    for plant 4:   US$8 per cloth
    for plant 5:   US$15 per cloth

- tins(18 kg) for plants 1,2,3 : US$0.60 each
- drums (150 kg) for plants 4,5 : US$7.50 each
- outside maintenance: 7 1/2 per cent of total equipment cost.

IV.2.5. Wages

The wages used in the economic analysis are based on prevailing rates in the Philippines and top of scale wages in India. The weighted average wage per worker is:

---

[1] IMF: International Financial Statistics (Washington, D.C.), various issues.

for plants 1, 2, 3: US$1,000 per year.

for plants 4 and 5: US$1,200 per year.

Plants 4 and 5 have higher wage costs than plants 1, 2, 3 because they require permanent employees who must receive legislated benefits. Plants 1, 2, and 3 can manage with daily labour.

IV.2.6. Product prices

The data on local prices of oil are more conflicting than those on raw material prices. However, they do tend to indicate that the average local price is slightly below world prices. On the other hand, there is no evidence that local prices of cake are below world market prices. Thus the following prices have been assumed:

- coconut oil: US$480 per metric tonne (10 percent below world market prices).

- groundnut oil: US$1,000 per metric tonne (10 percent below world market prices).

- oil cake: local price is assumed to be US$140 per metric tonne for virtually oil free (solvent extracted) cake. A premium of US$3 per tonne of cake for every additional 1 percent of oil content in cake is assumed for copra. A US$6 premium is assumed for groundnut cake. These assumptions are probably conservative. These premiums are limited to additional oil-in-cake up to a limit of 15 per cent, i.e. the upper unit prices for oil cake are:

US$140 + 15 x US$3 = US$185 for copra cake.

US$140 + 11 x US$6 = US$206 for groundnut cake as the maximum assumed oil content in cake is equal to 11 percent.

IV.3. Case study : two-ghani-power mill

This section provides an illustrative example of the application of the evaluation methodology described in the previous sections. The assumptions referred to in section IV.2 apply in this case study. The methodological framework is used to evaluate a two-ghani power mill for the extraction of oil respectively from copra and groundnuts.

IV.3.1. Two-Ghani (power) unit mill: oil extraction from copra

Step 1

- <u>input</u>: 560 kg of copra per day (151.2 tonnes per year).

- <u>organisation of production</u>: 1 shift per day, 270 working days
        per year.

- <u>output</u>: - oil: 0.57 x 151.2 = 86.2 tonnes/year
            - cake (18.5 percent oil-in-cake): 0.38 x 151.2 = 57.5
            tonnes/year.

Step 2: Fixed investment cost	US$
- land: 100 m^2 at US$5/m^2	500
- buildings (rural area): 36m^2 at US$80/m^2	2,880
- drying grounds: 20m^2 at US$10m/m^2	200
- equipment cost	4,000
- spares (5 percent of imported equipment cost)	175
Subtotal	7,755

Step 3: Working capital

2 days finished goods + 7 days raw material = 9 days

Therefore, working capital = 1.1 x 9 x 0.56 tonnes x
US$247/tonne   =   US$1.369.00

Step 4:   Total investment cost
    (2 + 3) = US$9,124.00

Step 5: Annual fixed investment costs	US$
- land: rental value at 10 per cent of cost	50.00
- buildings: 10 per cent pa, life 30 years therefore F = 9.427	305.50
- drying grounds: 10 per cent pa, life 6 years therefore F = 4.355	46.00
- equipment: 10 per cent pa, life 10 years therefore F = 6.145	651.00
- spares and maintenance: 7^1/2 per cent of US$4,000	300.00
Subtotal	1,352.50

Step 6:   Working capital annual cost

10 per cent interest on working capital
of US$1,369   US$ 136.90

Step 7:   Total annual fixed investment cost (5+6)   US$ 1,489.40

Step 8:   Annual variable costs   US$
    - raw material (copra): 151.2 tonnes at US$247/tonne 37,346.60
    - water   nil
    - electricity   nil
    - diesel fuel 2,430 litres at US$0.40 /litre   972.00
    - wood   nil
    - filter cloths: 24 at US$1.50 each   36.00
    - tin (18 kg): 4,800 at US$0.60 each   2,880.00
    - labour: 2 labourers at US$1,000 each   2,000.00
                                                     _____
                                       Subtotal   43,234.60
                                                     _____

Step 9:   Total annual costs (7+8)   US$44,724.00

Step 10:   Annual revenues
    - oil: 86.2 tonnes at US$480/ton   41,376.00
    - cake (15 percent oil-in-cake): 57.5 tonnes
      at US$185/tonne   10,637.50
                                 Total revenues   52,013.50

Step 11 : Annual gross profits
          US$52,013.50 - US$44.724   =US$ 7,289.50

IV.3.2. Two-ghani (Power) unit mill : oil extraction from groundnuts

Step 1:

    - input = 560 kg of groundnuts per day (151.2 tonnes per
    year).

    organisation of production: 1 shift per day,
      270 working days per year.

    output: - oil: 0.389 x 151.2 = 58.82 tonnes/year
             - cake (11 per cent oil-in-cake) = 0.541 x 151.2
                                      = 81.79 tonnes/year

Step 2   Fixed investment cost

Same as step 2 in section IV.3.1 =   US$ 7,755.00

Step 3   Working capital

2 days' finished goods + 7 days' raw material = 9 days
Therefore, working capital =
1.1 x 9 x 0.56 tonnes x US$470/ton =   US$ 2,605.00

Step 4   Total investment cost (2+3) =   US$10,360.00

Step 5   Annual fixed investment cost

Same as step 5 in section IV.3.1   US$ 1,352.50

Step 6   Annual working capital cost

10 per cent interest on working capital of US$2,605 US$ 260.50

Step 7   Annual total fixed investment cost (5+6)   US$ 1,613.00

Step 8   Annual variable costs   U.S.$
- raw material (groundnuts): 151.2 tonnes
  at US$470/tonne   71,064.00
- water   nil
- electricity   nil
- diesel fuel: 2,430 litres at US$0.40/litre   972.00
- wood   nil
- filter cloth: 24 at US$1.50 each   36.00
- tins (18 kg): 3,300 at US$0.60 each   1,980.00
- labour: 2 labourers at US$1,000 each   2,000.00
                                                      _____
                                    Subtotal   76,052.00.
                                                      _____

Step 9   Total annual costs (7+8) =   US$ 77,667.00

Step 10   Annual revenues
- oil 58.82 tonnes at US$1,000/ton   US$58,820.00
- cake (11 per cent oil-in-cake): 81.79 tons
  at US$185/ton   US$16,848.00

                                   _____
                              Total revenues   US$75,668.00

Step 11   Annual gross profits (Step 10 - Step 9) :
           US$75,668 - US$77,677 = - US$1,999.
     This particular unit is therefore operating at a loss.

IV.4. Economic comparison of alternative oil extraction technologies

     This section provides an economic comparison of the five oil extraction plants listed in section IV.2. Comparisons are made with respect to total investment costs, total annual costs and revenues and profitablity of the plants (annual profits and benefit-cost ratio).

     Findings from the economic comparison of alternative oil extraction technologies are provided as illustrative examples only and may not be used as a basis for technological choice for the following reasons. Firstly, as indicated in the previous section, these comparisons are based on a number of assumptions regarding factor prices, interest rate, output prices, etc. which will not generally apply to all developing countries. Secondly, these assumptions apply equally to the five technologies which are being compared. For example, the prices of raw materials is the same for the five types of plants. Yet, these prices should be lower for small-scale rural plants located near the oil seed growing areas than for large-scale urban plants, as the transport costs of raw materials should be considerably lower. Similarly, the oil produced by these plants is often marketed locally, thus minimising transport and marketing costs.

     Consequently, small-scale producers should enjoy a comparative advantage vis-à-vis large-scale producers with respect to these cost items. Other advantages may include low rural wages, low infrastructure costs, etc. Under these circumstances, interested readers should undertake their own evaluation of alternative oil extraction technologies and scales of production on the basis of local factor prices, transport costs, intended markets, etc., with a view to identifying the most suitable technology and scale of production.

A summary of the various cost and revenue items associated with the five oil extraction plants are provided in table IV.2 for copra and table IV.3 for groundnuts.

IV.4.1 Total investment costs

As shown in table IV.4 and IV.5 small-scale producers will need approximately between US$10,000 and US$45,000 if they were to invest in one of the three small-scale plants described in this memorandum (i.e. ghani mill, baby expeller, and small package expeller). Investments in medium or large expellers (1 to 6 million dollars) will generally be outside the financial reach of small-scale producers.

Investment costs per tonne of raw materials processed are as follows:

Plant No.	1	2	3	4	5
Copra	60.30	217.30	203.0	174.10	143.80
Groundnuts	68.50	225.50	212.00	223.20	217.40

The above table shows that investment costs per tonne of raw materials are by far the lowest for the ghani mill. Thus, small investors from countries suffering from shortages of capital funds (especially in the form of foreign currency) may find it easier to invest in these mills than in baby expellers or small package expellers.

IV.4.2 Annual costs and revenues

Table IV.4 (copra processing) shows that, for all technologies and scales associated with copra processing, revenues from the sale of oil only are lower than total annual costs. Thus, it is essential that small-scale producers be able to market the cake if they are to make a profit or break even. Otherwise, the sale price of oil must be higher than the international price of oil if the operation is to be profitable. However, it may be noted that the difference between costs and revenues (from oil) per tonne of raw materials processed is much lower for the power ghani mill than for the other small-scale plants (2 and 3).

Table IV.2
Summary data on the five oil extraction units

Copra processing

Items Plant designation →	Two-ghani mill 1	Baby expeller mill 2	Small package expeller 3	Medium expeller 4	Large expeller 5
Raw materials - (tonnes/hour)	0.070	0.050	.100	.800	4.0
Hours/day	8	8	8	20	24
Days/year	270	270	270	300	300
Raw materials - (Tonnes/year)	151.2	108	216	4 000	28 800
Tonnes oil/year	86.2	66.2	132.4	2 976	17 856
Tonnes cake/year	57.5	36.4	72.8	1 584	9 504
US$/tonne cake (Percentage oil-in-cake)	185 (18.5 %)	164 (8 %)	164 (8 %)	158 (6 %)	158 (6 %)
US$/Tonne oil	480	480	480	480	480
Land (m²) - cost US$	(100m²) 500	(90 m²) 450	(150 m²) 750	(1000 m²) 5 000	(2000 m²) 10 000
Buildings (m²)- cost US$	(36 m²) 2 880	(28.5 m²) 2 280	(50 m²) 4 000	(450 m²) 45 000	(1 050 m²) 105 000
Drying grounds (m²) - cost US$	(20 m²) 200	(16 m²) 160	(32 m²) 320	-	-
Equipment costs - US$	4 000	19 000	35 000	500 000	1 600 000
Cost spare parts - US$	175	600	1 600	25 000	80 000
Raw material cost -US$/year	37 746	26 676	53 352	1 185 600	7 113 600
Water (m³) - US$/year	-	-	-	(2 500 m³) 2 500	(10 000 m³) 10 000
Diesel fuel (litres) - US$/year	(2 430) 972	(2430) 972	-	(70 000) 28 000	(280 000) 112 000
Wood (tonnes) - US$/year	-	(8.1) 162	(16.2) 324	-	-
Electricity (kWh) - US$/year	-	-	(20 300) 1 827	(720 000) 64 800	(2 000 000) 180 000
Filter cloth (#) - US$/year	(24) 36	(18) 27	(30) 150	(200) 1 600	(500) 7 500
Labour (#) - US$/year	(2) 2000	(2) 2000	(3) 3 000	(16) 19 200	(48) 57 600
Packaging- US$/year	2 880	2 220	4 500	99 000	900 000
Interest on working capital - US$/year	137	98	217	26 083	234 748

Table IV.3
Summary data on the five oil extraction units

Groundnut processing

Items / Plant designation →	Two-ghani mill 1	Baby expeller mill 2	Small package expeller 3	Medium expeller 4	Large expeller 5
Raw materials - (tonnes/hour)	0.070	0.050	.100	.800	4.0
Hours/day	8	8	8	20	24
Days/year	270	270	270	300	300
Raw materials - (Tonnes/year)	151.2	108	216	4 000	28 800
Tonnes oil/year	58.82	44.28	88.56	2 035.2	11 211.2
Tonnes cake/year	81.79	58.32	116.64	2 524.8	15 148.8
US$/tonne cake (Percentage oil-in-cake)	206 (11 %)	185 (7.5 %)	185 (7.5 %)	170 (5 %)	170 (5 %)
US$/Tonne oil	1 000	1 000	1 000	1 000	1 000
Land (m²) - cost US$	(100 m²) 500	(90 m²) 450	(150 m²) 750	(1000 m²) 5 000	(2000 m²) 10 000
Buildings (m²)- cost US$	(36 m²) 2 800	(28.5 m²) 2 280	(50 m²) 4 000	(450 m²) 45 000	(1 050 m²) 105 000
Drying grounds (m²) - cost US$	(20 m²) 200	(16 m²) 160	(32 m²) 320	-	-
Equipment costs - US$	4 000	19 000	35 000	500 000	1 600 000
Cost spare parts - US$	175	600	1 600	25 000	80 000
Raw material cost -US$/year	71 064	50 760	101 520	2 256 000	13 536 000
Water (m³) - US$/year	-	-	-	(2 500 m³) 2 500	(10 000 m³) 10 000
Diesel fuel (litres) - US$/year	(2 430) 972	(2430) 972	-	(70 000) 28 000	(280 000) 112 000
Wood (tonnes) - US$/year	-	(8.1) 162	(16.2) 324	-	-
Electricity (kWh) - US$/year	-	-	(20 300) 1 827	(720 000) 64 800	(2 000 000) 180 000
Filter cloth (#) - US$/year	(24) 36	(18) 27	(30) 150	(200) 1 600	(500) 7 500
Labour (#) - US$/year	(2) 2000	(2) 2000	(3) 3 000	(16) 19 200	(48) 57 600
Packaging- US$/year	1 980	2 220	2 970	67 860	611 250
Interest on working capital - US$/year	260	186	414	49 632	446 688

Table IV.4

Economic comparison of coconut oil producing plants:

Plant No.	1	2	3	4	5
Type of Plant	Power ghani mill	Baby expeller mill	Small package expeller	Medium expeller	Large expeller
Input (copra) tonnes/year	151.2	108	216	4 800	28 800
Total investment cost (US$)	9 124	23 468	43 843	835 832	4 142 488
Total annual costs (US$)	44 724	36 996	72 264	1 535 295	8 957 946
Annual revenues from oil sales	41 376	31 776	63 552	1 428 480	8 570 880
Total annual revenues from oil and cake sales(US$)	52 013	37 745	75 491	1 678 752	10 072 512
Total annual gross profits(US$)	7 289	749	3 227	143 457	1 114 566
Benefit-cost ratio[1]	1 163	1 020	1 045	1 093	1 124
Output					
(a) Oil (tonnes/year)	86.2	66.2	132.4	2 976.0	17 856.0
(b) Cake (tonnes/year)	57.5	36.4	72.8	1584.0	9 504.0

[1] Benefit-cost ratio is equal to the total annual revenues divided by total annual costs. This ratio takes into consideration all benefits and costs, including depreciation costs.

Table IV.5:

Economic comparison of groundnut oil producing plants

Plant No.	1	2	3	4	5
Type of plant	Power Ghani mill	Baby expeller	Small package expeller	Medium expeller	Large expeller
Input (groundnuts) tonnes/year.	151.2	108	216	4 800	28 800
Total investment Cost (US $)	10 360	24 351	45 806	1 071 320	6 261 880
Total annual costs (US$)	77 665	61 168	119 098	2 598 104	15 303 536
Annual revenues from oil sales	58 820	44 280	88 560	2 035 200	12 211 200
Total annual revenues (US$) from oil and cake sales	75 668	55 069	110 138	2 464 416	14 786 496
Total annual gross profit(US$)	-1 997	-6 099	- 8 960	-133 688	-517 040
Benefit-cost ratio	.974	.900	.925	.948	.966
Output					
(a) Oil (tonnes/year)	58.82	44.28	88.56	2 035.2	12 211.2
(b) Cake (tonnes/year)	81.79	58.32	116.64	2 524.8	15 148.8

In the case of groundnuts (see table IV.5) revenues from the sale of both cake and oil are lower than total annual costs at all scales of production. These findings are the result of the assumed local and international prices of the raw materials, oil and cake. Special local circumstances should therefore be different from those assumed in this case study since a number of groundnut oil extraction plants do operate profitably (e.g. through government subsidies, high local prices for oil, low transport costs).

IV.4.3 Annual gross profits

Table IV.4 shows that all copra processing plants can be operated at a profit. However, the calculated benefit-cost ratios show that the power ghani mill (B/C = 1.163) is the most profitable at the indicated scale of production. It is followed by the large expeller (B/C = 1.124), the medium expeller (B/C = 1.093), the small package expeller mill (B/C = 1.045), and the baby expeller mill (B/C = 1.020). On the other hand, table IV.5 shows that all groundnut processing mills are not profitable. The two mills which are closest to the break-even point (i.e. no losses or gains) are the power ghani mill and the large expeller mill. They are followed successively by the medium expeller mill, the small package expeller mill and the baby expeller mill.

This case study shows that, given the adopted assumptions, the power ghani mill seems to be the most advantageous from the point of view of investment outlays per tonne of raw materials processed and the plant profitability. For those who may wish to produce at higher scales than allowed by the power ghani mill, the choice of the large expeller seems to be the most appropriate.

This case study also shows that the profitability of the various plants described in Chapter III are very much dependent on the prices of the raw materials and those of the oil and cake. It is therefore interesting to investigate the price levels which, given the selected technologies and costs of production, will make the selected plants break even. Such an analysis is carried out in the following section.

IV.5 Break-even prices of raw materials and oil

IV.5.1 Break-even prices of raw materials

It is of interest to know the maximum price of raw materials which should be paid by the producer in order to break even if the assumed processing costs (i.e. total annual costs excluding the cost of raw materials) and assumed market prices of oil and cake are to be valid. These maximum raw materials prices may be calculated according to the following steps:

(i) calculate processing costs by subtracting the cost of raw materials from total annual costs. For example, in the case of copra processing through the power-ghani mill, processing costs are equal to:

US$44,724 - US$37,346 = US$ 7,378.

(ii) calculate the total amount which may be paid for raw materials in order to break even by subtracting the processing costs from the total revenues. Thus, in the case of the power ghani mill, the maximum amount which may be paid for copra is equal to:

US$52,013 - US$7,378 = US$ 44,635.

(iii) finally, calculate the maximum price of raw materials in order to break even by dividing the cost of raw materials calculated in step (ii) by the total number of tonnes processed each year. Thus, the maximum price for copra is equal to:

US$44,635 ÷ 151.2 tonnes = US$295.21/ton.

Hence, given the assumed scale of production, processing costs and annual revenues, an oil producer may not pay more than US$295.21 per tonne of copra if he is to break-even. This estimated maximum price of copra is higher than the market price assumed in the case study (i.e. US$247/tonne), thus making copra processing by the ghani mill a particularly profitable operation.

Maximum prices of copra and groundnuts have been calculated for plants 1 to 5 as shown in table IV.6(a) for copra and table IV.7(a) for groundnuts. In the case of copra, the calculated maximum prices are higher than the one assumed in the case study (US$247/tonne). It is of interest to note that the largest difference between the calculated maximum price and the assumed market price is found for the power ghani mill (US$48.79 difference). Thus, producers which adopt this type of mill should be better protected against increases in the price of copra than if they had adopted the other types of oil extraction plants.

Table IV.6(a)

Maximum prices of copra

Plant No.	1	2	3	4	5
Total annual costs (excluding the cost of raw materials)($US)	7 378	10 320	18 912	349 695	1 844 346
Total annual revenues (US $)	52 013	37 745	75 491	1 678 752	10 072 432
Maximum price of copra to break even (US$/tonne)[1]	295.21	253.94	261.94	276.88	285.70

[1] The market price of copra assumed in the case study is equal to US$247/tonne.

Table IV.6 (b)

Minimum prices of copra oil

Plant No.	1	2	3	4	5
Total annual costs (US$)	44 724	36 996	72 264	1 535 295	8 957 946
By product (cake) annual revenues (US$)	10 637	5 969	11 939	250 272	1 501 632
Minimum price of copra oil to break even (US$/tonne)[2]	395.44	468.68	455.63	431.79	417.58

[2] The market price for copra oil assumed in the case study is equal to US$480/tonne.

In the case of groundnuts, (see table IV.7(a)), the calculated maximum prices for groundnuts are lower than the estimated market price of groundnuts (US$470/tonne) for the five plants. Thus, as already shown earlier, these plants may not run profitably if the assumptions made in the case study prove to be valid. Since a number of such plants are currently operating in a number of developing countries, it must be assumed that the prices of groundnuts and that of groundnut oil are different from the estimated international market prices. It should be noted that, as in the case of copra, the difference between the calculated maximum price and the estimated market price is lower for the power ghani mill than for the other plants.

IV.5.2 Break-even prices of oil

An analysis similar to that carried out for the price of raw materials may also be carried out in relation to the prices of copra and groundnut oil. In this case, we calculate the minimum price at which the oil must be sold in order to break even, given the assumed processing costs and market prices of the raw materials. This minimum price may be obtained as follows in the case of copra processing in the ghani mill:

- total annual costs - revenues from the sale of cake =
  US$44,724 - US$10,637 = US$34,087.
- The minimum price of oil to break even is therefore:
  US$34,087 ÷ 86.2 tonnes = $395.44/tonne.

Table IV.6(b) provides the minimum prices at which copra oil should be sold in order for the selected plants to break even. It may be seen that all calculated minimum prices are lower than the estimated market price of US$480/tonne. The minimum price of copra oil produced by the power ghani mill is by far the lowest minimum price. This finding provides additional evidence of the appropriateness of this mill should the assumptions made in the case study prove to be valid.

Table IV.7(b) provides the minimum prices at which groundnut oil should be sold in order for the selected plants to break even. It may be seen that the minimum prices for all plants are higher than the estimated market price (US$1,000 per ton). As in the case of copra oil, the calculated minimum prices of groundnut oil are closest to the market price for the power ghani mill and the large expeller mill.

Table IV.7(a)

Maximum prices of groundnuts

Plant No.	1	2	3	4	5
Total annual costs (excluding the cost of raw materials)(US$)	6 601	10 408	17 578	342 104	1 767 536
Total annual revenues (US$)	75 668	55 069	110 138	2 464 416	14 786 2496
Maximum price of groundnuts to break-even[1] (US$/tonne)	456.79	413.53	428.52	442.15	452.05

[1] The market price of groundnuts assumed in the case study is equal to US$470/tonne.

Table IV.7(b)

Minimum prices of groundnut oil

Plant No.	1	2	3	4	5
Total annual costs (US$)	77 665	61 168	119 098	2 598 104	15 303 536
By product (cake) annual revenues (US$)	16 848	10 789	21 578	429 216	2 575 296
Minimum price of groundnuts oil to break even (US$/tonne)[2]	1 033.95	1 137.73	1 101.17	1 065.69	1 042.34

[2] The market price for groundnuts oil assumed in the case study is equal to US$1,000/tonne.

IV.5.3 Concluding remarks on raw materials and oil prices

The choice of the appropriate oil extraction technology should, as shown in the previous section, improve the productivity of investments and labour in this sector and, therefore, the profitability of small-scale production units. However, the choice of technology affects mostly processing costs which represent a relatively small fraction (20 per cent to 40 per cent) of total production costs (i.e. processing costs plus raw materials costs). Thus, the market prices of raw materials and oil could have a significant impact on the profitability of an oil extraction mill. This is more so the case since the international market price of oil is not necessarily a function of total production costs. This may be explained by the fact that there are many other oil products which are good substitutes for groundnuts or copra oil (e.g. sunflower oil) and, therefore, the market price of these substitutes will necessarily affect the market price of groundnuts and copra oil. Thus, potential oil producers should carefully investigate the market of raw materials and the market where they intend to sell the produced oil. In many cases special circumstances may allow the profitable production of groundnuts and copra oil although the international prices of the raw materials and those of oil may militate against such production by small-scale producers. For example, the local prices of raw materials may be lower than the international prices because they do not include transport costs or transaction costs. Similarly, the produced oil may be competitive in the local market since similar transport and/or transaction costs may not also apply. In these cases, the choice of an appropriate oil extraction technology may play a decisive role by decreasing processing costs to the point where oil produced by small-scale units could become as or more competitive than oil marketed internationally.

CHAPTER V

SOCIO-ECONOMIC EVALUATION OF ALTERNATIVE
OIL EXTRACTION TECHNOLOGIES

Chapter IV provided guidelines for the estimation of the profitability of alternative oil extraction technologies as well as some illustrative examples of the application of these guidelines. The latter are mostly of interest to would-be or practising small-scale oil producers as the profitability of these technologies is considered exclusively from the point of view of the entrepreneur rather than that of society. On the other hand, public planners and project evaluators from industrial development agencies may also be interested in various socio-economic impacts of these technologies such as the generation of productive emloyment, the improvement of the country's balance of payments, rural industrialisation, etc. This chapter, therefore, analyses these various impacts through a comparison between a large-scale, capital-intensive plant and the small-scale plants (types 1 to 3) described in Chapter III (i.e. power ghani mill, baby expeller and small package expeller). The last section of this chapter suggests a few policy measures for the promotion of appropriate oil extraction technologies.

V.1 Employment impact

The generation of productive employment constitutes a major development objective for a large number of developing countries. Furthermore, improvement of the employment situation in rural areas is a prerequisite for the slowing down of the harmful rural to urban migration. Therefore, activities which favour the generation of rural employment should be of particular interest to public planners. The processing of groundnuts and copra in rural areas by small-scale, relatively labour-intensive mills may therefore be favoured over similar processing by large-scale plants located in urban areas.

In order to provide public planners with some indications of the impact on employment of alternative oil extraction technologies, comparison is made between a large-scale plant processing 28,800 tonnes per year of raw materials (groundnuts or copra) and small-scale plants (types 1 to 3) processing an equivalent amount of raw materials.

Table V.1 provides estimates of the employment that would be generated by the above plants. The number of small-scale mills needed to process 28,800 tonnes per year of raw materials varies from 133 mills (small package expeller) to 267 mills (baby expellers). Employment generated by these plants varies from 399 labourers (small package expeller) to 801 labourers (baby expeller), while the large-scale plant requires only 48 labourers. Thus, the small-scale plants generate from 7 times to over 17 times more labour than the large-scale plant for the same amount of processed raw material input. Furthermore, the small-scale plants would mostly generate rural employment while the large-scale plant would mostly generate limited urban employment. Thus, from an employment point of view the small-scale oil extraction mills are much more suitable than the large-scale mill, the baby expeller (type 1) being by far the most advantageous in terms of employment generation. However, the choice of oil extraction technology should not be solely based on the amount of employment generated. Other factors, which are analysed below, should also be considered.

V.2 Skills requirements

The availability of skilled labour often constitutes a constraint to the establishment of industrial units in developing countries. In many cases, these countries must depend on imported skills for the operation of large-scale, capital-intensive plants. Thus, technologies which would minimise such dependence should be favoured over others.

Table V.2 provides estimates of skills requirements for the three small-scale plants and the large-scale plant discussed in the previous section. It can be seen that plants 1 to 3 require unskilled and semi-skilled labour only. On the other hand, half of the labour of the large-scale plant is composed of semi-skilled and skilled labour, as well as labour with a high-school level education. Thus, from the point of view of skills requirements, the large scale plant may not be suitable under conditions prevailing in the majority of developing countries.

Table V.1

Employment generated by various oil extraction mills

Plants	Input (tonnes/year)	No. of plants required to process input	No. of employees required
1. Power ghani	28,800	190	570
2. Baby expeller	28,800	267	801
3. Small package expeller	28,800	133	399
4. Large-scale	28,800	1	48

Source: Tables IV.2 and IV.3. It is assumed that three employees are needed to run a small-scale plant (plants 1 to 3).

Table V.2

Skills requirements for selected oil extraction mills

Plant type	Small-scale plants			Large-scale plant
	1	2	3	4
Number of units	190	267	133	1
Unskilled	380	534	266	24
Semi-skilled	190	267	133	10
High-school level education	-	-	-	10
Skilled	-	-	-	4
TOTAL LABOUR	570	801	399	48

Source: Chapter IV (description of plants).

V.3 Investment and foreign exchange costs

The lack of sufficient capital and foreign exchange often constitutes a severe constraint on the development of industries in developing countries. Thus, technologies/scales of production which minimise the use of capital and foreign exchange should generally be favoured. In the case of oil extraction from groundnuts and copra, total investment costs required for the yearly processing of 28,800 tonnes of raw materials vary from US$1,473,450 for the power ghani (see table V.3 - Plant No. 1 - 190 power ghani units required) to US$6,004,800 for the baby expeller (267 units required to process 28,800 tonnes of raw materials). Total investments per tonne of coconut oil produced vary from US$89.8 for the power ghani mill to US$340.1 for the baby expeller. The corresponding figures for grounduts oil are respectively US$131.5 for the power ghani and US$508.5 for the baby expeller. Thus, from a purely investment point of view, the power ghani seems to be particularly advantageous. It may be noted that unit investment costs (i.e. per tonne of oil produced) are higher for the small-scale plants 2 (baby expeller) and 3 (small package expeller), than for the large-scale plant. Various reasons may explain this unexpected finding, including the lower productivity of equipment

Table V.3

Investment and foreign exchange costs (excluding working capital)

Plant	Input (tonne/ year)	No of plants required	Investment costs (US$)[1]				Foreign exchange costs (US$)[2]			
			Total	Per employee	Per tonne coconut oil produced	Per tonne ground- nut oil produced	Total	Per employee	Per tonne coconut oil produced	Per tonne ground- nut oil produced
1	28 800	190	1 473 450	2 585	89.8	131.5	665 000	1 167	40.50	59.40
2	28 800	267	6 004 830	7 496	340.10	508.50	3 204 000	4 000	181.50	271.30
3	28 800	133	5 542 110	13 890	313.90	469.40	4 256 000	10 666	241.10	360.40
4	28,800	1	1,795,000	37,395	100.50	147.00	1,400,000	29,166	78.40	114.60

Source: Chapter IV

---

[1] Working capital is not included.

[2] Imported equipment only

used in plants 2 and 3, various dis-economies of scale (e.g. in the use of land and buildings), etc. From an employment point of view, the power ghani mill is by far the most advantageous since total investment costs per employee amount to US$2,585 while these costs vary from US$7,496 for the baby expeller mill to US$37,395 for the large-scale plant.

Considering foreign exchange costs, the power ghani mill is also, by far, the most advantageous. Foreign exchange costs per tonne of oil produced by the ghani mill are respectively US$40.5 and US$59.4 for coconut oil and groundnut oil, while the corresponding costs for the other plants vary from US$78.4 to US$241.1 for coconut oil, and from US$114.6 to US$360.4 for groundnut oil. It may also be noted that foreign exchange costs per tonne of output are higher for the small-scale plants - except the power ghani mill - than for the large-scale plant. The same reasons as those provided for total investment costs may also explain the high foreign exchange costs associated with some of the small-scale plants. It should, however, be noted that the provided estimates of foreign exchange costs would not apply to all developing countries since some of the latter (e.g. India) may be able to produce some of the equipment which we assumed would be imported. Thus, foreign exchange costs for some of the small-scale mills could be much lower than those indicated in table V.3

V.4 Multiplier effects

The setting up of oil extraction mills could have a number of multiplier effects in the form of backward and forward linkages (e.g. production of oil containers, production of equipment for the oil mill). However, the importance of these multiplier effects would vary according to the adopted oil extraction technology and scale of production. In general, small-scale labour-intensive mills should generate larger multiplier effects than large-scale, capital-intensive mills for the following reasons:

> (1) A large proportion of the equipment used in small-scale mills may be produced locally, while most of the equipment used in large-scale mills must be imported.

> (2) Although oil containers (e.g. drums, tins) could in some cases be produced by local workshops for both small-scale and large-scale plants, the very large number of containers needed for centrally located large-scale plants may necessitate import of some of these containers if these workshops cannot ensure a steady supply of oil containers to these plants.

Additional multiplier effects should apply equally to large-scale and small-scale plants. These may include the processing of wastes (of coconut shells) into a number of products, such as activated charcoal, and the use of cake as feed for farm animals.

V.5 Energy and water requirements

High energy costs favour the use of technologies/scales of production which minimise the use of energy per unit of output. Furthermore, technologies which do not require conventional energy sources (e.g. electricity, fuel), may be easily applied in rural areas where these energy sources are not always available. Similarly, the lack of sufficient water in some areas of a country will favour technologies which do not require abundant water input.

Table V.4 provides estimates of various types of energy and water required by the three small-scale plants and the large-scale plant for the processing of 28,800 tonnes of raw materials per year. Two types of energy are considered in this table: electricity and diesel fuel. No electricity is needed for plants 1 and 2 (i.e. power ghani mill and baby expeller), while the electricity required per tonne of output is much higher for the small package expeller (plant 3) than for the large-scale plant. Diesel fuel is required for all plants with the exception of the small package expeller. However, fuel consumption by the power ghani mills is much lower than fuel consumption by the baby expeller but higher than that of the large-scale plant. Thus, from the point of view of energy requirements, the power ghani mill is particularly advantageous, considering that the large scale plant uses both diesel fuel and electricity.

Considering water inputs, the large-scale plant is the only one which makes use of water. Thus, the four small-scale plants are equally advantageous with respect to water utilisation.

The availability of electricity in rural areas and the unit prices of electricity and diesel fuel should determine which technologies are the most advantageous from the point of view of energy utilisation. For example, if very cheap electricity (e.g. produced by hydro-electric plants) is available in rural areas, the use of a small package expeller could be more advantageous than that of other types of oil extraction units which use imported diesel fuel. On the other hand, if rural areas are not electrified, the only alternative left to rural oil producers is to use mills powered by diesel fuel.

Table V.4

Annual energy and water requirements

Plant	Input (tonne/year)	No. of plants required	Electricity(KwH)			Diesel fuel (litres)			Water (m^3)		
			Total	Per tonne coconut oil	Per tonne groundnut oil	Total	Per tonne coconut oil	Per tonne groundnut oil	Total	Per tonne coconut oil	Per tonne groundnut oil
1	28 800	190	-	-	-	462 000	28.2	41.3	-	-	-
2	28 800	267	-	-	-	648 000	36.7	54.9	-	-	-
3	28 800	133	2 699 000	153.4	225.8	-	-	-	-	-	-
4	28 800	1	2 000 000	112.0	163.8	280 000	15.7	22.9	10 000	.56	.82

Source: Chapter IV.

V.6   Suggested policy measures for the promotion of suitable
       oil extraction technologies

A government wishing to promote suitable oil extraction technologies
may need to take into consideration the following factors:

Transport costs : Small-scale production units may be located in such a
way as to minimise transport costs associated with the transport of raw
materials and that of the outputs (i.e. oil and cake). Firstly, the
production units may be located close enough to the growing areas so as
to allow the transport of raw materials by low-cost transport means
available in rural areas (e.g. animal-drawn carts). Secondly, the
production of a small-scale unit is low enough to allow the marketing
of the output within a relatively short radius from the location of the
unit. On the other hand, processing by a large-scale plant, generally
located in main urban areas, will necessitate costly transport over
long distances for both the raw materials and the output. As transport
costs will be reflected in retail prices, consumers located far away
from these large-scale plants (especially rural consumers) will be
particularly penalised.

Quality of oil produced by small-scale units: If most of the local
demand is for unrefined oil, demand for refined oil by a minority of
the population may be satisfied through imports or, if it is large
enough, through the setting up of an oil refining plant. On the other
hand, if most of the oil is produced by large-scale plants equipped
with a refining unit, the tendency will be to produce and market
refined oil only. Thus, low-income consumers will be forced to pay for
the refining costs through high retail prices of refined oil.

Satisfaction of the basic needs of low-income groups:If a main
development objective pursued by the government is to increase
consumption by low-income groups, it is important that suitable
low-priced consumer goods be made available in sufficient quantities.
In the case of coconut and groundnut processing, the production of
unrefined oil by small-scale units should contribute to the fulfilment
of the basic needs objective for the following reasons. Firstly, as
stated above, low transport costs and the absence of refining should
yield lower retail prices than if the oil seeds were processed into
refined   oil   by   large-scale   plants   located   in   urban

areas. Secondly, packaging costs associated with oil production by small-scale units should also be relatively low as low-income consumers do not generally buy bottled oil: instead they bring their tins or bottles to the oil mill or retail shops. Furthermore, large drums for the transport of oil from the mill to retailers are generally re-utilised. Finally, local marketing of oil produced by small-scale mills minimises the necessity for intermediaries and should therefore result in a further lowering of retail prices.

Contribution to the achievement of major socio-economic objectives: As shown earlier in this chapter, small-scale oil extraction technologies generate substantially more employment than capital-intensive technologies, promote rural industrialisation and generate substantial backward and forward linkages.

Avoiding harmful monopolistic conditions: In countries where the processing and marketing of oil bearing seeds and fruits are undertaken by a very small number of large-scale producers, the latter can take advantage of their privileged position by offering low prices to the farmers for the raw materials, and selling the oil at the highest prices the market can bear. On the other hand, the existence of a large number of small-scale oil producers should create a much more competitive market for both the raw materials and the oil, and should therefore benefit both the farmers and the consumers. The higher prices paid to the farmers may also induce the latter to expand their production.

Given the above factors, governments could implement a number of policy measures which promote oil extraction techniques consonant with the adopted socio-economic objectives. The content of these measures will depend, to a large extent, on the extent to which the produced oil is intended for the local market only or for both local consumption and export.

If the produced oil is mostly for local consumption (i.e. the country produces limited amounts of raw materials), it may be in the interest of the country to implement various measures to ensure the efficient processing of raw materials by small-scale rural and urban extraction plants. As these plants produce unrefined oil only, a small oil refining unit may be established in order to satisfy demand for refined oil by various population groups. Furthermore, if a sufficiently large amount of cake is produced by the small-scale oil extraction units, this cake may be further processed in a solvent extraction plant should such processing be shown to be economically feasible. Such an approach may require the implementation of the following measures:

(1) The implementation of disuasive fiscal and monetary measures with a view to slowing down the adoption of inappropriate oil extraction technologies. These measures may be complemented by quotas or regulations regarding the import of oil processing equipment;

(2) The organisation of an efficient cake collection system from small-scale mills. Price incentives may be offered to millers who supply good quality cake for further processing;

(3) Introduction of improved oil extraction techniques, and training of labour and management with a view to improving productivity;

(4) Promotion of the local production of spare parts and equipment for small-scale mills. Training of technicians for repair and maintenance work;

The above measures may be complemented by others (e.g. promotion of production or service co-operatives, price incentives) depending on local conditions and circumstances.

If a country produces a surplus of copra and/or groundnut oil for export, exclusive reliance on small-scale extraction units may not always be feasible. Instead, countries may need to use both small-scale and large-scale units, especially if the oil is exported in a refined form. The large-scale units (large expellers and/or solvent extraction plants) may be supplied with both oil seeds and cake produced by small-scale units. The small-scale units may produce oil mostly for the local market, with any surplus of unrefined oil supplied to the large plants for further processing and exports. Careful planning of investments in the oil extraction sector will be needed in order to avoid a low capacity utilisation of the large-scale plants or the closing down of small-scale units for lack of raw materials (e.g. in cases where the capacity of large-scale plants is larger than what is needed to produce for exports. Thus, these plants may use raw materials intended for the small-scale mills and market the output locally). For example, the Government may limit, each year, the amount of oil seeds to be processed by large-scale plants depending on the amount of oil which can be exported, local demand for oil, and the availability of raw materials. As a rule, the processing of copra and/or groundnuts by small-scale mills for the local market should constitute a priority unless the seeking of foreign exchange constitutes a more important objective than the other objectives described in this section.

APPENDICES

APPENDIX I

Year	5%	6%	8%	10%	12%	14%	15%	16%	18%	20%	22%	24%	25%	26%	28%	30%	35%	40%
1	0.952	0.943	0.926	0.909	0.893	0.877	0.870	0.862	0.847	0.833	0.820	0.806	0.800	0.794	0.781	0.769	0.741	0.714
2	1.859	1.833	1.783	1.736	1.690	1.647	1.626	1.605	1.566	1.528	1.492	1.457	1.440	1.424	1.392	1.361	1.289	1.224
3	2.723	2.673	2.577	2.487	2.402	2.322	2.283	2.246	2.174	2.106	2.042	1.981	1.952	1.923	1.868	1.816	1.696	1.589
4	3.546	3.465	3.312	3.170	3.037	2.914	2.855	2.798	2.690	2.589	2.494	2.404	2.362	2.320	2.241	2.166	1.997	1.849
5	4.330	4.212	3.993	3.791	3.605	3.433	3.352	3.274	3.127	2.991	2.864	2.745	2.689	2.635	2.532	2.436	2.220	2.035
6	5.076	4.917	4.623	4.355	4.111	3.889	3.784	3.685	3.498	3.326	3.167	3.020	2.951	2.885	2.759	2.643	2.385	2.168
7	5.786	5.582	5.206	4.868	4.564	4.288	4.160	4.039	3.812	3.605	3.416	3.242	3.161	3.083	2.937	2.802	2.508	2.263
8	6.463	6.210	5.747	5.335	4.968	4.639	4.487	4.344	4.078	3.837	3.619	3.421	3.329	3.241	3.076	2.925	2.598	2.331
9	7.108	6.802	6.247	5.759	5.328	4.946	4.772	4.607	4.303	4.031	3.786	3.566	3.463	3.366	3.184	3.019	2.665	2.379
10	7.722	7.360	6.710	6.145	5.650	5.216	5.019	4.833	4.494	4.192	3.923	3.682	3.571	3.465	3.269	3.092	2.715	2.414
11	8.306	7.887	7.139	6.495	5.938	5.453	5.234	5.029	4.656	4.327	4.035	3.776	3.656	3.544	3.335	3.147	2.752	2.438
12	8.863	8.384	7.536	6.814	6.194	5.660	5.421	5.197	4.793	4.439	4.127	3.851	3.725	3.606	3.387	3.190	2.779	2.456
13	9.394	8.853	7.904	7.103	6.424	5.842	5.583	5.342	4.910	4.533	4.203	3.912	3.780	3.656	3.427	3.223	2.799	2.468
14	9.899	9.295	8.244	7.367	6.628	6.002	5.724	5.468	5.008	4.611	4.265	3.962	3.824	3.695	3.459	3.249	2.814	2.477
15	10.380	9.712	8.559	7.606	6.811	6.142	5.847	5.575	5.092	4.675	4.315	4.001	3.859	3.726	3.483	3.268	2.825	2.484
16	10.838	10.106	8.851	7.824	6.974	6.265	5.954	5.669	5.162	4.730	4.357	4.033	3.887	3.751	3.503	3.283	2.834	2.489
17	11.274	10.477	9.122	8.022	7.120	6.373	6.047	5.749	5.222	4.775	4.391	4.059	3.910	3.771	3.518	3.295	2.840	2.492
18	11.690	10.828	9.372	8.201	7.250	6.467	6.128	5.818	5.273	4.812	4.419	4.080	3.928	3.786	3.529	3.304	2.844	2.494
19	12.085	11.158	9.604	8.365	7.366	6.550	6.198	5.877	5.316	4.844	4.442	4.097	3.942	3.799	3.539	3.311	2.848	2.496
20	12.462	11.470	9.818	8.514	7.469	6.623	6.259	5.929	5.353	4.870	4.460	4.110	3.954	3.808	3.546	3.316	2.850	2.497
21	12.821	11.764	10.017	8.649	7.562	6.687	6.312	5.973	5.384	4.891	4.476	4.121	3.963	3.816	3.551	3.320	2.852	2.498
22	13.163	12.042	10.201	8.772	7.645	6.743	6.359	6.011	5.410	4.909	4.488	4.130	3.970	3.822	3.556	3.323	2.853	2.498
23	13.489	12.303	10.371	8.883	7.718	6.792	6.399	6.044	5.432	4.925	4.499	4.137	3.976	3.827	3.559	3.325	2.854	2.499
24	13.799	12.550	10.529	8.985	7.784	6.835	6.434	6.073	5.451	4.937	4.507	4.143	3.981	3.831	3.562	3.327	2.855	2.499
25	14.094	12.783	10.675	9.077	7.843	6.873	6.464	6.097	5.467	4.948	4.514	4.147	3.985	3.834	3.564	3.329	2.856	2.499

Present worth of an annuity factor

How much 1 received or paid annually for X years is worth today

APPENDIX II

EQUIPMENT MANUFACTURERS

The following list of manufacturers of oil extractor equipment is far from complete. Other manufacturers, in both developed and developing countries, most probably exist which could supply the various pieces of equipment described in this memorandum. However, information at hand at the time the memorandum was published did not allow an exhaustive coverage of equipment suppliers. Would-be or practising oil producers are therefore urged to identify, in the trade catalogues (see, for example Appendix IV) or from importers of equipment, other manufacturers who may supply the type of equipment they are looking for. It must also be emphasised that the equipment suppliers referred to in this Appendix do not imply their endorsement by the International Labour Office, and any failure to mention a particular supplier in connection with the technologies described in this volume is not a sign of disapproval.

Belgium

De Smet,   Solvent extraction equipment
265 avenue Prince Baudouin,
Edegem-Antwerp

Brazil

Masiero Industrial S.A.   Pre-press expellers,
P.O. Box 218-219,   solvent extraction equipment
Jan Sao Paulo,

China

China National Machinery Import   Small oil expellers
and Export Corporation
Shandong Branch,
82 Fan Hsiu Road,
Tsingtao

Federal Republic of Germany

Mathias Reinartz,   Decorticators, seed cleaners,
Neuss   expellers

Freid Krupp,   Full press expellers
Harburger Eisen-Und Bronzewerke,   Pre-press expellers
Hamburg 90   Solvent extraction equipment

Christiansen & Meyer,   Solvent extraction equipment
Hamburg-Harburn Aussenmulhenweg 10

Extraktiontechnik Geselschaft für Olmuhleneirichtungen, 200 Hamburg, 13 Werderstrasse	Solvent extraction equipment
Lurgi, D-6000 Frankfurt/Main 2	Solvent extraction equipment

France

Etablissement A. Olier, Clermont-Ferrand	Decorticators, seed cleaners, expellers , presses, solvent extraction plants

India

Hindsons Pvt. Ltd., Patiala, Punjab	Foot-operated groundnut shellers
The Punjab Oil Expeller Co., Patel Marg, Ghaziabad	Small expellers, fitter presses
S.P. Engineering Corp., P.O. Box No. 218, Kanpur	Small and medium scale expellers, decorticators, seed cleaners, small scale extraction plants
Kirloskar Brothers Ltd., Poona	Decorticators and seed cleaners
Swastik Expeller Industries, Bombay	Baby oil expellers, small-scale expellers
Kamal Engineering Corporation, Nagpur	Table model expeller
Dandekar Bros., Sangli, Maharashtra	Hand-operated groundnut decorticators
United Engineering (Eastern) Corporation, 22 Biplabi Rash Behari, Bose Road, Calcutta 1	Decorticators, seed cleaners, expellers, small-scale extractions plants
United Oil Mill Machinery and Spares, D-298 Defence Colony, New Delhi	Decorticators, seed cleaners, expellers, small-scale extraction plants
Premier Engineering Co., Cochin	Indirect hot-air copra driers
Numex Engineers, P.O. Box 820, Bombay	Small-scale oil extraction plants
Steel construction Co. Pvt, Bangalore	Seed cleaners

Israel

H.L.S. Ltd Industrial Engineering Company, P.O. Box 193, Petah-Tikuah	Solvent extraction equipment

Italy

Construzzioni Meccaniche Bernardini, C.M.B., Via Petronella, 0040 Pomezia (Roma)	Solvent extraction equipment
Fratelli Gianazza S.P.A., Via Le Cardona 78/84, 20025 Legano	Solvent extraction equipment

Japan

Chuo Baeki Goshi Kaiska (CECOCO), P.O. Box 8, Ibraki City, Osaka Pref.	Small-scale oil extraction plants, groundnut decorticators.

Luxembourg

Usine de Wecker Luxembourg	Presses

Malawi

Agrimal (Malawi) Ltd. P.O. Box 143, Blantyre	Groundnut shellers

Netherlands

Stork-Apparatenboun BV, P.O. Box 3007, Amsterdam	Complete oil mills

Switzerland

Buhler-Miag Ltd., Uzwil	Decorticators, seed cleaners

Tanzania

Ubongo Farm Implements, P.O. Box 2669, Dar es Salaam	Groundnut decorticators

United Kingdom

Messrs. Rose, Downs and
  Thompsons Ltd.,
Cannon Street,
Hull

Large scale presses, and solvent
extraction plants

Harrap Wilkinson Ltd.,
N. Phoebe Street,
Salford

Automatic groundnut decorticating
machines

R. Hunt and Co. Ltd.,
Atlas Works,
Earls Colne,
Colchester,
Essex CO6 2EP

Groundnut decorticator

United States

V.D. Anderson Company,
Cleveland,
Ohio

Various pieces of oil extraction
equipment and entire expeller plants

Anderson IBEC,
19699 Progress Drive,
Strongfield,
Ohio 44136

Full press expellers
Pre press expellers
Small oil expellers

French Oil Mill Machinery Company,
Pique,
Ohio 45356

Solvent extraction equipment

Full press, pre-press expellers

Dravo Corporation, Chemical Plants
Div.
One Oliver Plaza,
Pittsburgh,
Pa 15222

Solvent extraction equipment

Bauer Brothers Co.,
Springfield,
Ohio

Various pieces of oil extraction
equipment and entire expeller plants

APPENDIX III

LIST OF INSTITUTIONS AND AGENCIES INVOLVED IN OIL EXTRACTION

1.   FAO, Fats and Oils Section,
      Commodities and Trade Division,
      Economic and Social Department,
      Via delle Terme di Caracalla
      00100 Rome,
      Italy

2.   African Groundnut Council
      P.O. Box 3025,
      Lagos,
      Nigeria

3.   Asian Coconut Community,
      2nd Floor, Pantja Niaga Building,
      94-96 Djalan Kramat Raya,
      Djakarta,
      Indonesia

4.   Oil Technological Research
        Institute (OTRI),
      Anantapur,
      Andhra Pradesh,
      India

5.   Philippine Coconut Oil Producers Association,
      Singson Building,
      Room 309,
      Manila,
      Philippines

6.   Coconut Research Institute of Ceylon,
      Bandirippura Estate,
      Lunuwila,
      Sri Lanka

7.   Tropical Products Institute,
      Oil Palm Advisory Bureau,
      56/62 Gray's Inn Road,
      London WC1,
      United Kingdom

8.   National Peanut Council,
      1120 Connecticut Avenue,
      Washington, D.C. 20036 ,
      United States

9.   Department of Transport, Works and Supply,
      P.O. Box 1108,
      Boroko,
      Papua New Guinea

APPENDIX IV

<u>LIST OF DIRECTORIES WHICH PROVIDE INFORMATION ON EQUIPMENT MANUFACTURERS</u>

Australasian Manufacturers' Directory,
    Manufacturer Publishing Co. Pty Ltd.,
    Elizabeth and Hill Streets,
    North Sydney,
    N.S.W. Australia

Thomas' Register of American Manufacturers,
    Thomas Publishing Co.,
    461 8th Avenue,
    New York, N.Y. 1001,
    (United States)

Europ. Production,
    ABC der Deutschen Wirtschaft,
    Berliner Allee 8,
    61 Darmstadt,
    Federal Republic of Germany

Kompass Directories,
    Kompass International AG,
    Neuhausstrasse 4,
    8044,
    Zurich,
    Switzerland

APPENDIX V

SELECTED LIST OF JOURNALS ON OIL EXTRACTION

Ceylon Coconut Quarterly,
Publ:Coconut Research Institute,
    Bandirippura Estate,
    Lunuwilla,
    Sri Lanka

Tropical Products Quarterly,
Publ:Commonwealth Secretariat,
    Printing Section,
    Pall Mall,
    London, SW1,
    United Kingdom

Coconut Bulletin,
Publ:Directorate of Coconut Development,
    Ministry of Food and Agriculture,
    Government of India,
    Ernakulam 1,
    India

APPENDIX VI

SELECTED BIBLIOGRAPHY

Achaya, K.T: Appropriate technology for production and processing of oils and fats, Paper presented at International Forum on Appropriate Industrial Technology (New Delhi, UNIDO, 1978).

Aten, A., Manni, M., and Cooke, F.C.: Copra processing in rural industries FAO, Agricultural Development Paper No. 63 (Rome, FAO, 1958).

Boyd, John: A buyer's guide to low-cost agricultural implements (London, Intermediate Technology Publications Ltd., 1976).

De, S.S. and Cornelius, J.A.: Technology of production of edible flours and protein products from groundnuts, Agricultural Services Bulletin No. 10 (Rome, FAO, 1971).

Godin, V.G. and Spensley, P.C.: Oil and oilseeds, Crop and Product Digests No. 1 (London, TPI, 1971).

Grimwood, Brian E.:Coconut palm products (Rome, FAO, 1976).

Hagenmaier, Robert D.: Coconut aqueous processing (Cebu City, Philippines, San Carlos Publications, 1980).

Rao, P.V.S: A study of village oil industry in India (Lucknow, Appropriate Technology Development Association, 1980).

Thampan, P.K.: The coconut palm and its products (Cochin, India, Green Villa Publishing House, 1975).

Thieme, J.G.: Coconut oil processing (Rome, FAO, 1968).

Tropical Products Institute: An economic study of lauric oilseed processing (London, 1973).

Tropical Products Institute: An economic evaluation of the wet coconut process developed at the Tropical Products Institute (London, 1973).

Tropical Products Institute: Crop and product digest No. 1 (London, 1971).

Tropical Products Institute: Development of a wet coconut process designed to extract protein and oil from fresh coconut (London, 1973).

UNIDO: Appropriate industrial technology for oils and fats (New York, United Nations, 1979).

UNIDO: Coconut processing technology information documents, Part 1 to 7, Coconut harvesting and copra production (Vienna, 1980), limited distribution.

UNIDO: Coconut Processing Technology Information Documents, Part 2 of 7, Coconut oil extraction (Vienna, 1980), limited distribution.

UNIDO: Guidelines for the establishment and operation of vegetable oil factories (New York, United Nations, 1977).

QUESTIONNAIRE

1. Full name...

2. Address...

3. Profession (check the appropriate case)

    (a) Small-scale oil producer ....☐

    (b) Official from the Agriculture Ministry ....☐

    (c) Extension officer ....☐

    (d) Staff member of a training institution ....☐
         If yes, specify ....
         ....
         ....

    (e) University staff member ....☐

    (f) Staff member of a technology institution ....
         If yes, specify ....☐
         ....
         ....

    (g) Staff member of a financial institution ....☐

    (h) Official of a Planning Ministry or Agency ....☐

    (i) International technical assistance expert ....☐

    (j) Other ....☐
         If yes, specify ....
         ....

4. From where did you get a copy of the memorandum?
    Specify if given free or bought ....
    ....

5. Did the memorandum help you achieve the following
    (Check the appropriate case(s))

    (a) Learn about oil extraction techniques you
         were not aware of ....☐

    (b) Estimate unit production costs for various
         scales of production/technologies ....☐

    (c) Improve your current production technique ....☐

(d) Cut down operating costs ....□

(e) Improve the quality of produced oil ....□

(f) Decide which technology to adopt for a new
oil extraction unit ....□

(g) If a government employee, to formulate new
policies and measures in favour of the
small-scale oil extraction sector ....□

(h) If an employee of a financial institution,
to assess a request for a loan for the
establishment of a small-scale oil
extraction plant ....□

(i) If a trainer or extension officer, to use the
memorandum as a supplementary training material ...□

(j) If an international expert, to better advise
counterparts on oil extraction techniques ....□

(k) Indicate any other help provided by this
memorandum ....□
....
....
....

6. Is the memorandum detailed enough in terms of:
    - Technical aspects .... Yes   No

    - Costing information .... Yes   No

    - Information on socio-economic impact .... Yes   No

If some of the answers are "No", please indicate why
below or on a separate sheet of paper ....
....
....
....
....
....
....
....
....
....

7. How might this memorandum be improved if a second
    edition were to be published? ....
    ....
    ....
    ....
    ....
    ....
    ....
    ....
    ....
    ....

8. Please send the questionnaire, duly completed, to:
    Technology and Employment Branch,
    International Labour Office,
    CH-1211 GENEVA 22   (Switzerland)

9. In case you need additonal information on some of
    the issues covered by this memorandum, the ILO and
    FAO wil do their best to provide the requested
    information.